Transatlantic Energy Futures

Strategic Perspectives on Energy Security, Climate Change, and New Technologies in Europe and the United States

Edited by
David Koranyi

Center for Transatlantic Relations
Paul H. Nitze School of Advanced International Studies
Johns Hopkins University

Koranyi, David, *Transatlantic Energy Futures: Strategic Perspectives on Energy Security, Climate Change, and New Technologies in Europe and the United States*

Washington, DC: Center for Transatlantic Relations, 2011.

Center for Transatlantic Relations
The Paul H. Nitze School of Advanced International Studies
The Johns Hopkins University
1717 Massachusetts Ave., NW, Suite 525
Washington, DC 20036
Tel: (202) 663-5880
Fax (202) 663-5879
Email: transatlantic@jhu.edu
http://transatlantic.sais-jhu.edu

ISBN 0-9848544-0-1
ISBN 978-0-9848544-0-0

Table of Contents

Introduction

The world of energy is in upheaval. "It is a time of unprecedented uncertainties," as the International Energy Agency put it in its 2010 World Energy Outlook report. Economic recovery after the crisis is still fragile. The hunger for energy resources gnaws at rapidly emerging economies; energy poverty is rampant. 1.4 billion people lack access to electricity; New York City uses as much energy as 800 million people in Sub-Saharan Africa. Energy markets are in flux as producer and consumer positions shift dramatically. Even as Russia has overtaken Saudi Arabia as the world's largest oil producer, the United States has surpassed Russia as the world's largest producer of natural gas. Global energy markets are increasingly interconnected and interdependent, yet global energy governance is still in its infancy. Climate change challenges loom ever larger, while a global agreement on how to tackle them seems distant. The Fukushima-Daiichi nuclear accident has cast doubt over the future of nuclear power, and the revolutions in the Middle East, spawned by millions of people looking for dignity and a better life, have upended governments and erased complacent assumptions about energy supply and demand.

With these developments in mind, CTR Fellow David Koranyi has gathered an impressive set of authors to explore how Europe and the United States can grapple with the energy questions of today and tomorrow. What drives policy decisions in Washington and in the capitals of U.S. states, or in Brussels and in European countries? What will define their energy mixes in the future? What are the similarities and differences, convergences and divergences among the various energy sectors in Europe and the United States? What should be done to facilitate transatlantic cooperation in the field of energy from a political, diplomatic, institutional, commercial, regulatory and financial perspective? Is a transatlantic energy alliance desirable? Is it possible? What could be the goals, scope, shape and influence of such an alliance? Their conclusions are well worth reading.

This project unfolded as part of a larger initiative conducted by the Center for Transatlantic Relations, which is examining common challenges facing the United States and Europe. We want to thank the European Union for its support of this effort, and to thank many colleagues who participated in the deliberations and meetings that produced this book.

We would also like to thank our colleagues at the Center for Transatlantic Relations for their help and good cheer throughout this project, and Peggy Irvine and Peter Lindeman for working with us on the many details related to the production of the book.

Our authors express their own views, and do not necessarily reflect views of any institution or government.

Daniel S. Hamilton
Director, Center for Transatlantic Relations
Paul H. Nitze School of Advanced International Studies
Johns Hopkins University

Towards a Transatlantic Energy Alliance: Prospects for EU-U.S. Cooperation in Fighting Climate Change and Promoting Energy Security and New Technologies

David Koranyi

The year is 2035. A tornado has just hit Washington, D.C., followed by severe flooding; a regular feature lately as global warming has led to extreme weather conditions in the United States and all over the world. The President of the EU looks out from the window of her plane about to land at Reagan Airport, and marvels at the craftsmanship of the engineers who built the 60 meter high giant dam stretching hundreds of miles along the East Coast. "What was the price, like 3 trillion dollars? How long will they last?" she wonders. 3-4 years maximum. Next time she comes, she will have to fly two more hours to Atlanta, which has been designated as the new U.S. capital. She just left a crisis summit in Brussels, where European leaders had to decide whether to take in another 20 million refugees from Africa, where the humanitarian situation had become intolerable after years of draught and the ensuing military conflicts. "We knew what was coming. We had every card in our hands. Yet we were complacent. The world looked upon us for leadership. Yet we failed to act. There is no reason in political madness." she thinks, as the plane lands and the motorcade leaves for the White House.

She greets the U.S. President with a chilly smile as they proceed to talks in the Roosevelt Room. It will be a long day, petty wrangling about how to distribute the remaining assets of NATO, a once mighty military organization that has been dismantled recently over pressure by the Chinese. Beijing made this as a precondition, she recalls—for providing financial assistance and technological expertise to modernize their economy. "If only we were wise enough back then"—she thinks, as the doors close...

The year is 2035. The U.S. President has just landed in Brussels—the futuristic but still rainy capital city of Europe. He was here only two months ago to celebrate the 20th anniversary of the signing of the Global Climate Change Law at the UNFCCC COP21 conference in the city—the "last minute deal to save the Earth," as headlines reported at the time. "We could all be grateful to the Europeans"—he speculates. "Without their sometimes annoyingly stubborn nagging, it could have ended up very differently." He thinks proudly though of the role of his nation. "Americans can always be counted on to do the right thing—after they exhausted all other options," Churchill once quipped. How true, he thinks. "We woke up late, but not too late. Without our genuine creativity and witty innovations, humankind would have failed." As the car rolls in front of the refurbished Justus Lipsius building - that except for the trees on top looks more like a spaceship with all the solar panels—he spots a new meter-like gadget in the presidential car—an American brand using European technology. "What's this?" he asks his aide and she is equally perplexed. "Mr. President, that is a smart meter of sorts," the driver jumps in. "This is of course an electric car, but as a backup it still has a fossil fuel engine. Would we have been using that, we would have pay a charge by the kilometer. Remember, we are in Europe." The President nods as he recalls the uphill struggles to change the wasteful energy-consumption habits of his own people. It took painful measures: high taxes and tough regulations—any politician's nightmare. But it was worth it as the voters - eventually—recognized.

"We could do this via Skype, just like in a family" he thinks, but remembers that they will have a guest today. His fellow European leader welcomes him with a warm hug. They jointly walk towards the red carpet to meet the Chinese President. He is here to discuss terms and conditions for green technology transfer for Chinese companies...

A glimpse into two fundamentally different futures. Only 24 years from now. The transatlantic community stands at a fork in the road. Will we take the one leading to chaos and despair or the one to a bright, sustainable future? On face value it seems like an obvious choice. Yet it is anything but that. The undeniable reality of climate change sets us against a highly complex and interconnected set of challenges that are political, economic, technical, social and cultural in nature. To solve them requires unprecedented global political cooperation and institutional coordination, a previously unseen spirit of solidarity and sacrifice, maximum human ingenuity and fundamental changes in actual or aspired lifestyles around the planet.

Multiple crises complicate this already daunting task. We have to face five watershed challenges simultaneously. The political deadlock in Washington and the euro crisis in Brussels draws attention away from energy and climate change threats. The economic crisis conceals the gravity of the situation as energy demand growth and emissions drop temporarily in the developed world. The ensuing social crisis, coupled with limited political capital, inhibits leaders from asking even reasonable sacrifices. The financial crisis constrains the ability to invest in technology, innovation and infrastructure[1]. All this culminates in a perceived crisis of the 'Western model' of shared values based on human dignity, democracy, rule of law and market economy.

The climate crisis sets a very tight deadline for changing policy course and managing a successful transformation to a low-carbon economy. Add to this the ever-increasing demand from emerging and developing countries, the uncertainty of the energy markets in flux, the fierce debates over the sustainability of nuclear energy, natural gas or biofuels, and the geopolitical tensions over energy resources, and one has a recipe for disaster.

Yet there is hope. Amidst these challenges, three positive trends emerge that hold the promise of a better world. The first is technological development. We have a good chance to deploy technologies in time that are capable of stopping global warming—such as carbon capture and storage, new battery technologies, smart grids, solar panels or advanced biofuels. The bottom line is that many of these still require significant Research, Development and Demonstration (RD&D) to achieve technological breakthroughs that would make them cheaper and more efficient. Then there are the unforeseen, potentially game-changing energy sources that could radically rewrite the playbook. Technology and improving energy efficiency together could ensure that energy demand growth will remain manageable.

---

[1] The IEA puts the investments needed simply to meet projected increases in energy demand by 2035 to $40 trillion, one-third of that in the developed world, largely in the EU and the U.S.

The second positive trend is what Jeremy Rifkin calls the 'democratization of energy.'[2] As Rifkin writes:

> *"The new era will bring with it a reorganization of power relationships across every level of society. While the fossil fuel-based First and Second Industrial Revolutions scaled vertically and favored centralized, top-down organizational structures operating in markets, the Third Industrial Revolution is organized nodally, scales laterally, and favors distributed and collaborative business practices that work most effectively in networks. The "democratization of energy" has profound implications for how we orchestrate the entirety of human life in the coming century ... In the coming era, hundreds of millions of people will produce their own green energy in their homes, offices, and factories and share it with each other in an "energy Internet," just like we now create and share information online. The democratization of energy will bring with it a fundamental reordering of human relationships, impacting the very way we conduct business, govern society, educate our children, and engage in civic life."*[3]

Clean energy technologies could empower people both politically and economically, especially those 1.4 billion still without basic access to electricity. Moreover, democratization of energy already started with the "cleanest" of fossil fuels: natural gas. The discovery of huge unconventional gas reserves all over the world, as opposed to the conventional resources limited to only a handful of countries, could fundamentally change relationships between producers and consumers in the coming decade. It could dampen volatility, reduce the influence of petro-states and, as their rents narrow, loosen the grip of authoritarian regimes on their own people. As a local, decentralized source of fuel, unconventional gas is evolving as an important balancing factor.

The third trend is the emergence of a sensible mainstream clean energy movement that is globalized in nature and involves a colorful

---

[2] Jeremy Rifkin, "The 'Democratization Of Energy' Will Change Everything," available at http://www.huffingtonpost.com/2011/09/26/jeremy-rifkin-democratization-of-energy-green-technology_n_980222.html.

[3] Jeremy Rifkin, *The Third Industrial Revolution: How Lateral Power is Transforming Energy, the Economy, and the World* (London: Palgrave Macmillan 2011).

coalition among a growing number of people from different walks of life: academia, the corporate world, the media, and the NGO community—be they environment-conscious citizens who engage in bottom-up activities already to the tune of Rifkin's democratization vision or interest-led entrepreneurs sensing a good business opportunity. Their influence on international and national policymakers cannot be underestimated and has already produced some tangible results.

Core Dilemmas

No matter how high the stakes, there is no foolproof guarantee either that the transformation to a low-carbon economy will inevitably take place, or that it will be timely and orderly.

First, as of this writing,[4] successful conclusion of global climate change negotiations seems rather distant. Virtually all major players in the global community who have put their mind around the problem of climate change acknowledge the need for action. What we face is a classic game theory prisoner's dilemma, where countries want emission reductions but prefer that someone else take on the burden. Emerging economies have a point in emphasizing their right to develop. Therefore, enhancing the sense of justice coupled with peer pressure is the key to adopting a successful climate change regime. A new approach that puts per capita emissions, rather than historical issues, at the center of the debate would help address both problems and could result in an efficient, flexible, transparent and just system.[5]

Second, even if the global framework is in place, its implementation, monitoring and sanctioning will be extremely complex and challenging from an international governance perspective, and peppered with hiccups and failures. Institution-building, conditionality and the right incentives, such as linking climate change with trade,[6] are also crucial to overcome the free-rider problem.

---

[4] November 2011.

[5] One example for this approach: "International climate change negotiations: Key lessons and next steps"; Smith School of Enterprise and the Environment, University of Oxford, July 2011. Available at http://www.smithschool.ox.ac.uk/wp-content/uploads/2011/03/Climate-Negotiations-report_Final.pdf.

Third, the transformation to a clean energy economy will be long, costly and involve deep and initially unpopular changes in lifestyle in the Western world. Especially in the U.S., higher energy (especially electricity and gasoline) prices should be accepted as part of reality, as only real economic costs can change consumption patterns and prevailing attitudes towards energy production and usage.

Fourth, managing the transformation to a low-carbon economy will entail tough policy calls in government and tricky investment decisions in the private sector. It is clear that the transformation cannot happen overnight. There are considerable uncertainties regarding the speed and extent at which new clean energy technologies can be deployed. Therefore 'bridges' are needed to a low-carbon future. Conventional and unconventional natural gas can play an important role in the global energy mix—gas could reduce energy dependence in the U.S. and in Europe and become a bridge fuel to a low-carbon future. However, anxieties about a lock-in effect are not without some justification. Moreover, there could be fierce competition for the role of bridge fuel and preferred base-load power generation, primarily between natural gas and nuclear. While the former is branded as the cleanest fossil fuel, as it emits approximately 50% less than coal, and the discovery of widely available unconventional assets could considerably ease angst over access constraints, the IEA assesses that "*an increased share of natural gas in the global energy mix is far from enough on its own to put us on a carbon emissions path consistent with an average global temperature rise of no more than 2 degrees Celsius.*"[7] On the other hand, zero-emission nuclear energy struggles with public acceptance issues and financial difficulties, understandably so after the Fukushima accident in March 2011.

Fifth, even though energy investments will have to be taken up largely by the private sector, ultimately public policy choices will need to determine the preferred path by setting forth a stable and predictable regulatory environment and incentives system. It is clear that the private sector and markets alone cannot trigger the transformation

[6] Such as imposing trade tariffs on countries that are unwilling to regulate carbon emissions.

[7] *Are We Entering a Golden Age of Gas?*—IEA Special Report, 2011 http://www.iea.org/weo/docs/weo2011/WEO2011_GoldenAgeofGasReport.pdf.

within the limited time available. Therefore strategic vision, policy signals, regulatory incentives and public funding are crucial to trigger technological change and ensure that the benefits of the "green revolution" are shared by everyone. This does not mean that governments should enter into micromanagement, pick winners and losers, nor that they should shoulder the bulk of costs. Governments, however, have a special responsibility in two regards: to set a firm policy course towards a low-carbon future by providing regulatory and economic incentives; and to ensure a level playing field for all clean energy resources. Legitimate concerns over competitiveness, both in the developed and emerging economies, must also be taken into account in designing a comprehensive system to reduce carbon emissions.

Sixth, current and potential resource scarcities[8] can further complicate the picture, such as rare earth elements that are needed for the production of renewable energy-generating equipment.

Seventh, it is a delicate task to find the right balance between rivalry and cooperation on the global stage in terms of energy technologies and innovation. A healthy dose of competition and national/regional support schemes must ensure that the private sector keeps up the momentum in developing and deploying technologies eventually with a return on their investments. At the same time synergies must be be tapped and major projects such as nuclear fusion pursued jointly. Technology transfer is crucial to ensure that developing countries that are less well off could join the 'green revolution' early on.

What Brings Us Together...

Transatlantic cooperation is key to addressing all the above challenges and dilemmas. Due to a number of reasons, the transatlantic partners are well positioned to provide answers jointly.

To begin with, transatlantic cooperation on energy has a rich history, a decent track record and a good basis upon which to build. It

---

[8] Christof van Agt, "Buying Time: Energy and the Art of Sustainable Advancement in Transatlantic Relations, " in Daniel Hamilton and Kurt Volker, eds., *Transatlantic 2020: A Tale of Four Futures* (Washington, DC: Center for Transatlantic Relations, John Hopkins University, October 2011), pp. 257-284.

picked up after the first oil crisis in 1973-74 and led to the establishment of the International Energy Agency (IEA). In the 1980s the transatlantic partners somewhat differed in their views on core energy security issues and in their responses to challenges, such as the role of Russia in providing oil and natural gas to Europe. Nonetheless, transatlantic cooperation again intensified in the 1990s and 2000s on various issues, such as oil and gas pipelines,[9] energy efficiency, RD&D cooperation, carbon capture and storage projects, smart grids, and energy storage. This culminated in the establishment of the EU-U.S. Energy Council in November 2009, which testified to the recognition of energy as an issue of strategic importance and of great potential in transatlantic cooperation.

The transatlantic partners share strategic interests in maintaining and improving the effectiveness of a global governance system that is norm-based, rule-based, and inclusive, and that ensures the security of the U.S and the EU. Moreover, the EU and the U.S. have an exceptionally strong incentive—exacerbated by the financial and economic crisis—to reinforce existing cooperation and to share burdens by pooling resources. In times of austerity and shrinking budgets, identifying and exploiting synergies and avoiding duplications is a must.

The transatlantic community is uniquely positioned to develop technology, leverage financing, and share experiences in legislative and regulatory developments that are necessary to advance clean energy technologies. As pluralist democracies, the EU and the U.S. are best positioned to profit from the 'democratization of energy.' Innovation, initiative, subsidiarity and self-governance, decentralized decision-making system, management of interconnectivity, co-dependencies and market integration— all these skills, which will be required to be successful in the new era, are deeply ingrained in our societies.

Finally we face common threats and challenges closely linked to energy issues, such as the proliferation of nuclear weapons, a resurgent Russia, an unstable Middle East or China's insatiable appetite for resources and its repercussions around the globe.

---

[9] Such as the Baku-Tibilisi-Ceyhan oil pipeline or the planned Nabucco gas pipeline.

...and What Drives Us Apart

Critical factors of divergence cannot be discounted either, as they have an almost equally strong pull. Differing climate change perceptions and the lack of U.S. commitment and action is extremely dangerous, as it alienates Europeans, both policymakers and the wider public alike. These differences, if not solved, could drive a wedge for decades between the partners, undermine trust, create a value gap and hinder cooperation not only in climate change and energy issues but in all other aspects as well.

There is in fact a chance that U.S. and European energy markets could largely decouple in coming years, due in part to differences regarding the need to tackle climate change, and in part to diverging geopolitical and domestic trends. The U.S. has edged closer to self-sufficiency with respect to fossil fuels, with the extensive development of its vast unconventional gas resources and increasing reliance on Canadian oil sands. This could lead to a more isolationist stance in U.S. policy. Meanwhile unconventional gas faces mixed reactions in Europe; the EU, for example, plans to shun oil shales and tar sands in its impending Fuel Quality Directive. Friction in transatlantic perceptions on energy security and divergences over preferred courses of action are real dangers that must be addressed head on.

Towards a Transatlantic Energy Alliance

The systemic transformation of the world of energy, triggered by climate change and powered by new technologies, will likely cause the reorganization of our societies. The benefits and pitfalls of transatlantic cooperation are beyond doubt. Renewing the transatlantic community's leadership is essential to lead the world to a sustainable, low-carbon future. Transatlantic cooperation can contribute to provide secure and affordable energy to people in the EU and the U.S., foster economic prosperity and create jobs. Current cooperation on a wide range of subjects is encouraging but inadequate. What we need is a new impetus, genuine political will, adequate resources and enhanced cooperation to advance a transatlantic green economy. Joint efforts in addressing climate change, innovation and investment into clean energy technologies, risk sharing and cost reduction, joint RD&D and harmonized energy diplomacy must be the cornerstones of a Transatlantic Energy Alliance.

A Transatlantic Energy Alliance is desirable and feasible, but not self-evident. Climate change and energy cooperation will be the litmus test of converging or diverging European and American norms, values and interests in the 21st century. We have to bridge our differences and we have to do that quickly in order to remain in the driving seat. To amend Robert Kagan's famous line, Americans may be from Mars and Europeans from Venus, but we shall all soon need to move to some other planet if we do not adjust course.

* * *

Transatlantic Energy Futures endeavors to give you a taste of the intricate and multifaceted energy challenges facing our communities. It aims to do so with a strong conviction in the enduring prominence and necessity of the transatlantic partnership.

Given the vast expanse of the subject, we could not possibly cover all aspects of transatlantic energy cooperation. The book consciously focuses more on the supply side and runs through almost all fuel types.[10] Global energy governance, climate change, carbon trading, innovation, smart grids, and supply and transit security are cross-cutting issues of utter importance that deserved separate chapters. Mixing the horizontal approach with sectoral studies offers a better picture of present and future trends in the transatlantic energy arena. The demand side is equally important, of course, and also ripe for enhanced cooperation between the EU and the U.S.—a topic, perhaps, for a subsequent book.

I deliberately chose a proactive approach with the purpose of contributing to the intensive thinking on both sides of the Atlantic over these pressing issues. I have asked the authors to be ambitious in offering advice to present and future policymakers. They did so graciously, enabling me to come up with a set of recommendations that are by no means intended to be comprehensive but certainly meant to be helpful.

[10]Major omissions are hydro, wind and coal (oil is covered in Kirsten Westphal's excellent chapter 1), the latter by no means a fuel of the past, given its prevalence in global primary energy supplies and in electricity generation.

I am grateful to the Johns Hopkins SAIS Center for Transatlantic Relations, to directors Dan Hamilton and Kurt Volker for this opportunity, for the support of the European Commission and for the devotion and solid work of my fellow authors in putting together this book. It was a privilege to work with such distinguished and competent experts on a subject that is highly relevant to the future of transatlantic relations.

Executive Summary

David Koranyi

Kirsten Westphal sets the scene with a comprehensive overview of the uncertain geopolitical and macroeconomic environment of the global energy system. She outlines three global energy challenges that demand urgent, simultaneous and multi-level political action: ensuring energy security for every country; minimizing the effects of energy systems on the environment and the climate; and providing access to modern forms of energy to all people. The sheer magnitude of the triple challenge is daunting due to its urgent, global and systemic nature.

Westphal recalls that the International Energy Agency's "estimates of the world's total endowment of economically exploitable fossil fuels and hydroelectric, uranium and renewable energy resources indicate that they are more than sufficient to meet the projected increase in consumption to 2035."[1] The question is rather how to manage the transformation, whether the necessary investments will happen in time and in the right place.

Recent years have been characterized by extreme volatility in prices and markets. Meanwhile, international energy governance, which could dampen this volatility is fragmented and weak. Given the complex nature of relations among actors, the interdependence of energy systems, and interconnections among the energy sectors, as well as the interplay along the whole value-added energy chain, governing or even steering the energy transition is a Herculean task.

Climate change exacerbates these problems. After high hopes for a U.S. policy shift in 2008, Mark Olsthoorn argues that we are back to square one. The fallout between the EU and the U.S. was quite considerable at the 2008 Copenhagen climate change meeting, and the two are again estranged partners with regard to global climate change

[1] IEA, *World Energy Outlook 2010*, p. 117.

and clean energy issues. Further widening of perceptions and a gap in actions would, in the medium-term, be disruptive to the entire transatlantic relationship. Olsthoorn contends that the tables will eventually turn also in the U.S. and that Washington will eventually subscribe to a comprehensive global climate framework. Until then, the EU should build momentum by engaging with regional partners in the U.S. and national partners elsewhere; and to prepare by setting boundary conditions for fast, cost-effective development of low-carbon energy system through increased and improved energy RD&D spending and collaboration, policy evaluation, building a skilled workforce and market integration.

Until the tables turn, however, transatlantic clashes over climate policy might even intensify. Pál Belényesi discusses the European carbon trading system after 2012 and its implications to the U.S., arguing that the EU ETS is modelled on previous U.S. approaches, later abandoned, but that the EU has chosen a significantly different way of incentivizing the participants. Many argue that the system creates perverse incentives for CO_2 intensive plants to remain in operation in order to receive free allocations. In addition, firms might invest in and operate more carbon-intensive technologies if they anticipate that future allocations of allowances will be proportional.

There are many questions left to be answered. How does the U.S. approach the third trading period in the EU, where U.S.-based companies are also clearly involved in the must-reduce, must-comply, must-buy/sell trading scheme, as air traffic joins the scheme? Will the U.S. fight EU legislation at international aviation forums? Does the U.S. plan to counteract with a federal scheme that would eventually effect European players too? Or will it address the issue in a cooperative way coordinate with the EU and the developing world while reducing its emissions?

Mihaela Carstei writes about the role of public and private financing in energy RD&D, arguing that public energy RD&D should be targeted to riskier and more innovative technologies in the initial stages of development, including prototyping and demonstration projects, so that it has real added value and avoids crowding out private investments. All energy technologies must compete in the marketplace in terms of cost, reliability and ability to attract capital, but the recent financial crisis and economic downturn, coupled with a number of key energy sector developments, has dramatically changed the investment landscape. The current global financial crisis has created significant constraints from which the U.S. and European energy industries are not immune. Money is exceptionally tight, and will likely remain so for several years. As a result, overall momentum and public and private investments in new research and development projects are slowing. This will dramatically inhibit the pace at which the U.S. and European energy industries can be transformed.

Carstei strongly believes that the transatlantic community is uniquely positioned to develop the technology, financing, and legislative and regulatory developments necessary to advance energy security. Further, there is growing recognition throughout the transatlantic community that the energy sector needs to be radically transformed in order to achieve energy security and to reduce the impact of climate change while ensuring economic prosperity. Thus, pursuing an interrelated set of strategic objectives will ultimately require similar structural changes in the way energy is supplied and used throughout the transatlantic community. Moreover, the transformation of the energy sector will entail considerable analysis of new policies, regulations and enforcement mechanisms, as well as the deployment of many new technologies that are promising, but not yet proven. The foundation for transatlantic energy security will be the establishment of common, compatible and complementary strategies that allow for enhanced innovation.

Nikolas Foster discusses how smart grids could be a test case of such cooperation on joint innovation and technology deployment. While initial motivations regarding smart grids are different in the United States and Europe, several overlaps in technological and regulatory challenges call for closer cooperation. Both the United States and Europe are using the same technology and are tackling similar problems on the road to smart grids. Both could benefit from lessons learned so far, especially given the substantial investments in the United States following the 2009 stimulus funding.

Foster emphasizes that as the United States is examining its path towards smart grids, countries in the EU are confronted with similar policy choices that warrant a look at the benefits of transatlantic smart grid cooperation. Comparing experiences can help U.S. states with pending deregulation and European countries, which are beginning the prescribed liberalization process, with formulation and application of best practices. Deregulation is especially interesting to examine in light of ongoing smart grid technology deployments. Questions such as how deregulation affects smart grid deployment could be answered by looking at the experiences and challenges taking place on the other continent. Examining each other's experiences, especially in times of constrained resources, will benefit both the United States and Europe. Mutual learning and further cooperation in smart grid deployment brings the promise of advancing the transatlantic relationship as both continents adapt to a new era of advanced technology, constrained resources and rising global demand for energy.

Reinis Aboltins also outlines a promising future for transatlantic cooperation in renewable energies, as both the EU and the U.S. constitute huge markets for renewable energy technologies as well as renewable energy resources. Aboltins notes that it is both the similarities and the differences in energy production and consumption in the U.S. and the EU that provide enormous potential for mutually beneficial synergies. He underlines that in terms of investment, the renewable energy sector is definitely a good market: renewable energy investment worldwide rose to $211 billion in 2010. While the majority of this funding goes to finance large-scale deployment projects rather than R&D or early commercialization activities, the level of financing indicates that there is great interest in renewable and clean energy technologies as good investments. Another shift is also notable: with investments of $72 billion in utility-scale renewable energy projects and companies, for the first time developing countries have spent more than developed economies on renewable energy. China has led this surge; its nearly $50 billion invested in renewable energy in 2010 makes it by far the largest source of, and destination for, clean energy investment globally. This clearly shows from where future competition in renewable energy production will be coming. Europe and the U.S. have to bear this in mind and act.

Among renewables, solar power has enormous potential, though it also faces serious challenges, as Michael Stanton-Geddes explains. Throughout human history, engineers and researchers have attempted to utilize the sun's energy by turning solar irradiation into a medium that provides heating, cooling, and lighting services. However, today direct solar energy accounts for only about 0.1% of total primary energy supply worldwide, and the international community still speaks of solar energy as a "new energy" to power the world tomorrow.

Stanton-Geddes argues that the most stubborn challenges are socio-political, not technological. The solar industry is on the verge of achieving its long-predicted potential, with economic, development and environmental benefits. The European Union and the United States followed remarkably similar paths towards the development of solar energy policy, regardless of substantial differences in the details. Both countries face the problem of competition from countries with lower labor costs and environmental standards. The U.S. and EU could use this area to reinvigorate cooperation and indicate that the transatlantic relationship today is about more than simply stopping financial market contagion; it is also about taking on constructive projects. Making use of the momentum as well as newest technological advances, the EU and U.S. should adopt a new joint approach to solar energy research, investigate better subsidy policy designs, and facilitate international trade in solar energy equipment.

Are all renewables really that 'green'? The sustainability and desirability of some have been questioned time and again. Tamás Kenessey tells us why serious concerns have been raised about the sustainability of biofuels, and that a number of serious social and ethical issues also have to be addressed, such as the link between growing biofuel production and rising food prices. The rapid growth in the consumption of first- generation biofuels, largely driven by policies in the United States and the European Union, led to serious contradictions. The use of edible feedstock contributed to increasing food prices, and greenhouse gas emission savings lagged behind expectations.

Kenessey presents biofuels as essential for the decarbonization of the transport sector, despite all their present flaws. Advanced technologies go beyond the food versus fuel debate and provide considerable greenhouse gas reductions compared to petroleum. To achieve the dual objective of improving energy security and reducing greenhouse gas emissions, the United States and the European Union have to revise their current biofuel policies. A coordinated action during this revision would equally benefit the transatlantic partners and the entire biofuel industry. The U.S. and the EU have to increase significantly their R&D expenditures on next generation biofuels and advanced agricultural practices. They have to implement sound sustainability criteria and promote the adoption of reliable international certification schemes. They should cooperate closely when setting biofuel mandates and targets, taking into account the international implications of their domestic measures.

David Koranyi recalls that a year ago the energy world was abuzz with talk of a nuclear renaissance. Since the Fukushima nuclear crisis, however, there is an ongoing rethinking of nuclear energy policy across the world. In Europe, the disaster accelerated the denuclearization agenda in Germany and Spain, phasing out of nuclear plans in Switzerland and Belgium, abandoning nuclear energy development plans in Italy, and increased uncertainty around nuclear development plans elsewhere in the continent. In the United States there seems to be political consensus around and less public debate about nuclear energy; nevertheless, the future of nuclear industry is hard to predict due to various unknowns in the economic and political circumstances.

Koranyi argues that in order to effectively combat climate change and to reduce energy poverty around the world, nuclear power is indispensable. However the constraints on governmental and private sector decisions in the U.S. and Europe triggered by the Fukushima accident may result in protracted political and regulatory processes, which could further raise costs for nuclear development to unbearable levels. That would effectively kill the development prospects of the nuclear industry in the U.S. and Europe for the foreseeable future, prolong the extended use of fossil fuels as base-load power and prevent the large-scale reduction of greenhouse gas emissions. The United States and Europe should play a pivotal role in advancing the efficient and responsible use of nuclear energy, the enhancement of safety standards and the further development of the international regulatory and institutional setup.

Competition for base-load power generation between nuclear and natural gas could be the epic story of coming years. Maximilian Kuhn and Frank Umbach argue that natural gas is an attractive "transition fuel" towards a low-emission global energy mix and that it will be a critical element in the fundamental climate, economic and political calculations made by all major economies, primarily as a transition fuel for a renewable energy mix for decades to come. Natural gas is and will remain an essential component of the energy mix for a variety of reasons: massively increased global and regional established gas reserves; competitive economic costs; a relatively favorable carbon footprint when properly developed and transported; and a natural synergy with the large-scale development of intermittent electricity sources. Most recently, gas has provided a partial substitute for the loss or delay of (new) nuclear power generation after the Fukushima disaster. Natural gas is increasingly seen as a viable source of energy due to improvements in technology, changed regulatory frameworks, and the fact that gas is seen as the 'greenest' fossil fuel.

Kuhn and Umbach point out that unconventional gas has become the new 'elephant in the room' with global geopolitical implications that have caused a chain reaction: European gas prices are being renegotiated and revised. It has also caused an average of 15% of Gazprom's supplies to be delinked from oil-indexation in 2010. But the implications are greater still: relatively cheap and abundant gas, along with the carbon advantage of gas, makes nuclear power and coal relatively more expensive than currently assumed. Indeed, making gas a major transition fuel through 2030 will help renewable energy efforts to reduce emissions, at a lower cost in order to mitigate the impact of climate change

Kornél Andzsans-Balogh elaborates on natural gas and its role in transatlantic energy security. The largest natural gas consumers of the world, the United States and the European Union, are on the brink of having a single common natural gas market pricing. This phenomenon is a new addition to transatlantic relations and requires active support from policymakers to become reality.

Andzsans-Balogh underlines the key energy security aspects of natural gas in the transatlantic context. The largest currently known conventional natural gas reserves are to be found along the axis of northwest Siberia down to Qatar. While the United States is independent from the resources of this axis, the European Union views the South-Siberian-Persian axis as a potential source region for its current pipeline imports, which are depleting over the midterm, and as an alternative to currently dominant Russian supplies. As the EU taps into the region's resources, the Caspian region's geopolitical importance will increase. In global natural gas supply, Qatar's role will be balanced out by new capacities in Australia. However, Doha will remain a key supplier for European LNG supplies. Europe has not been able to extend its capabilities beyond the tools of soft power. Therefore, its increasing reliance on distant resources will require support in the framework of transatlantic cooperation.

Geopolitics and Russia looms large on the agenda when it comes to transatlantic energy security. András Deák elaborates on energy relations between Russia and the West, highlighting the Russian energy industry's peculiarities and their effects on energy security in Europe and the U.S. The Russian energy security sector has a strong autarkic perspective. Apart from its early 19th century beginnings, it developed within an almost exclusive national context; cooperation with foreign majors has been much more limited than has been the case with other producers. Moreover, Russia became a great power long before it started its energy exports. Unlike Saudi Arabia or Iran, for which oil was almost the single reason to become a foreign policy actor, Russia has to integrate its global energy significance into a broader set of external relations and perceptions. This makes Russia a peculiar actor. It has relatively less experience with having foreign concessions on its soil, but has high expectations regarding its status and leverage on world general and energy matters.

Deák argues, however, that Russia is now approaching the limits of its autarkic industry patterns. It will have to establish a new, more cooperative policy in oil production if it wants to secure its production level in the long term. It has to show a more interactive behavior with gas exports if it wants to keep its markets in the midterm. Thirty years ago similar constraints and the failure of the Soviet leadership to manage them played a considerable part in the collapse of oil production and in the fall of the Soviet Union. The situation has changed since then. The Cold War is over, but Russian leaders remember and fear a similar outcome, while energy exports play a crucial role in current Russian stability. All this provides a chance to set the foundation for a moderate, if often uneasy, partnership with Russia in the field of energy.

Ukraine is another key component of European energy security, as Taras Kuzio explains. After two decades as an independent state, corruption has become an integral part of Ukrainian society, business and politics. The energy sector has been the most tempting for corrupt elites from all political groups and regions as the greatest rents can be extracted from it that give incumbents greater advantage in politics.

Kuzio warns that high levels of corruption in Ukraine's energy sector have ramifications upon transatlantic and European security as they lead to the prioritization of corruption and short-term interests over reforms; the export of corruption to western European NATO and EU members; instability in energy relations between Russia and Ukraine, which could lead to a repeat of the 2006 and 2009 crises; and prioritization of personal over national interests, as seen in the imprisonment of Julia Tymoshenko, which threatens to derails Ukraine's European integration and leads to its possible geopolitical re-orientation towards Russia and the CIS.

Recommendations

The recommendations below are drawn mainly from the studies in this book but in their entirety reflect the views of the editor only and are by no means intended to be comprehensive. They invoke and build upon excellent earlier reports on the prospect of transatlantic energy and climate change cooperation by the Johns Hopkins SAIS Center for Transatlantic Relations,[2] the Atlantic Council,[3] the Clingendael International Energy Program,[4] the Center for Strategic and International Studies[5] and the Heinrich Böll Foundation.[6]

[2] http://transatlantic.sais-jhu.edu/transatlantic-topics/energy-security.htm; http://transatlantic.sais-jhu.edu/sebin/c/s/us-eu_report_final.pdf.

[3] http://www.acus.org/files/publication_pdfs/65/AtlanticCouncil-USEUEnergy-Rev4.pdf.

[4] http://www.clingendael.nl/publications/2011/20111000_transatlantic_vanagt.pdf.

[5] http://www.acus.org/files/publication_pdfs/523/EnergySecurityReport.pdf.

[6] http://www.boell.org/downloads/HBF_FinalReport_Sharing_Solutions.pdf.

I. Horizontal Issues

Climate Change

- A comprehensive, legally binding global agreement on climate change in the UNFCCC framework is essential to limit the increase in global temperature below 2 degrees Celsius by at least halving global greenhouse gas emissions by 2050 compared to 1990 levels and based on per capita emission levels.
- The U.S. must recognize the stakes and join the EU in introducing a price on carbon, preferably by a cap-and-trade scheme modeled on the EU Emissions Trading System (ETS) and by subscribing to global efforts by pushing for and signing up to a comprehensive and compulsory global effort to fight climate change in the UNFCCC framework. Joint EU and U.S. leadership is key to garner international support for the agreement.
- Until a firm U.S. commitment on the above, there are four prime objectives for transatlantic collaboration:
    - The EU should keep advocating for a global framework of binding emissions reductions and build momentum by engaging with regional partners in the U.S. and national partners elsewhere.
    - The EU should integrate the ETS with regional carbon trading schemes in the U.S. through common verification mechanisms and harmonized reduction targets. That could create a transatlantic carbon market and enhance competition for cost-effective reduction technologies and identify regulatory bottlenecks that could be problematic later on.
    - The ETS must be reinforced and the feasibility and desirability of introducing a European-wide carbon floor price must be examined.
    - The Anti-ETS Bill in the U.S.Congress must urgently be revoked, before it causes further complications for the aviation industry on both sides of the Atlantic.

Clean Technologies

- Clean energy development is one of the most promising area for transatlantic cooperation in coming decades. The EU and the U.S. must be at the vanguard of clean energy technologies and set the boundary conditions for fast, cost-effective development of a low-carbon energy system through increased and improved energy RD&D spending and collaboration, policy evaluation, building a skilled workforce and market integration.
- A well thought-out combination of 'technology push' and 'market pull' instruments is needed to create an efficient clean energy technology innovation and commercialization engine. Key elements:
    - Priority technologies should be identified through a systematic assessment, taking into account the importance of the technology toward the national energy goals and the comparative advantages.
    - Instruments should be tailored to a priority technology (or technology group) and the development stage a technology is in. The further it has progressed the more the market forces should be leveraged.
    - Ensure strong competition between companies within a family of technologies.
    - Closely and carefully monitor against milestones to minimize waste.
- Public policies should generate sufficient investment security for private funders to invest in early application of emerging technologies and implement market pull policies that allow economies of scale to bring down cost. Loan guarantees, feed-in tariffs and performance standards are instruments that can be successful in these phases if designed appropriately. Harmonization of regulations related to technology export or exchange in the EU and the U.S. must be promoted.
- Education is crucial in changing lifestyles to rationalize energy consumption and building a large enough workforce

with the right skills (primarily in science, technology, engineering and mathematics) is key to ensure the smooth transition to a low-carbon economy. This is an essential area of transatlantic cooperation. Creating a network of training centers and universities and the establishment of a Transatlantic Clean Energy Academy could be useful in that respect.

- The EU and the U.S. must boost energy innovation by creating a U.S.-EU Clean Energy Bank and a Transatlantic Energy Innovation Fund.[7]
- The transatlantic partners must look for synergies and enhanced cooperation between the U.S. ARPA-Energy program and the EU's Strategic Energy Technology Plan (SET). The EU and the U.S. should coordinate on sector-specific initiatives such as joint carbon capture and sequestration projects.
- A transatlantic clean energy partnership is vital to encourage the spread of low-carbon technologies in developing countries. That would entail reciprocal efforts to open markets for foreign investments and optimal, innovation-effective protection of intellectual property rights. Binding emissions reductions contingent on technology assistance from developed countries.
- The EU and the U.S. must create common transatlantic intellectual property and standards policies that promote vigorous competition. Policies to reward and protect innovation are equally important. The transatlantic community should aim to develop complementary competition and property laws that encourage innovation while enhancing economic efficiency and consumer welfare. Furthermore, policymakers should establish a framework that allows for the rights to intellectual property to be shared appropriately and emphasizes the commitment of governments on both sides of the Atlantic to ensuring consistent and effective enforcement.
- All clean technologies must be tested against a sustainability criteria to ensure that the overall benefits outweigh the costs.

[7] Along the lines of the December 2009 report—http://transatlantic.sais-jhu.edu/sebin/c/s/us-eu_report_final.pdf.

The U.S. and the EU should coordinate their actions on the development of joint and sound sustainability criteria that link financial support to GHG performance and incentivize the efficient, economical and sustainable use of natural resources.

Market Access and Integration

- The EU and the U.S should be at the forefront of creating a level playing field for clean energy technologies to allow alternative energy options to compete on their true socio-economic merits among others by introducing carbon cap and trade systems and dismantling fossil fuel subsidies as agreed in the G20 framework.
- The EU and the U.S. must speed up the formation of domestic and international clean energy markets by providing a stable regulatory framework and strong incentives such as renewable portfolio standards and loan guarantees.
- The EU and the U.S. must create a competitive market for transatlantic energy innovation. A key policy mechanism to encourage and promote innovation in the development and deployment of energy technologies is the removal of trade and investment barriers. A transatlantic commitment to promote open investment policies, avoid new trade restrictions, and aim to eliminate existing barriers is essential to create an attractive environment for much needed capital flows.
- The EU and the U.S. must push for reciprocity of investments vis-á-vis third players, like Russia and China.
- The transatlantic partners must work to enhance transparency and competition in energy markets and cross-border investments.
- The EU must vigorously push for market liberalization by the full implantation of the Third Energy Package thus creating a single, interconnected European energy market.
- It is of utmost importance for the EU and the U.S. to ensure that competition is not distorted, with special regard to cartelization in the regional and global gas markets.

- In the long term the EU and the U.S. must promote global competition rules for energy markets to reduce cartelization and prices, cripple the disruptive ability of irresponsible players on the market, enhance security of supplies and promote improving efficiency.

Financing

- As many of the technologies needed to transform our economies to run on clean energy are not yet available, sustained and substantial research and development is required to bring these technologies to market in time and at an affordable price. The capital requirements are considerably higher than what the private sector can support so there is a role for government support.
- Pooling resources to share costs and risks shall be a key element in transatlantic cooperation as public funding will be constrained by the large public debts in the aftermath of the financial and economic crisis. Enhanced transatlantic cooperation is an imperative to vigorously pursue basic scientific research, reduce the costs of emerging technologies through accelerated learning; share costs for research and development on technologies that are expected to provide public benefits; and promote the development of innovative capabilities at home and abroad to promote long-term competitiveness.
- The EU and the U.S. must develop and support transatlantic public research initiatives that are tightly linked to the private sector and acknowledge and respond to non-technical challenges to innovation.
- Policymakers on both sides of the Atlantic should create effective regulatory frameworks and long-term incentives for public-private partnerships. These would necessarily include: developing new models for the funding, commercialization and deployment of new innovative advances; funding for precompetitive research that can aid in bringing technologies their commercial stage; building human capital and knowledge stocks by increasing the availability of transatlantic scholarships and other financial support for the science fields; sup-

port the use of performance based technology regulation and standards; promote the creation of a robust information technology infrastructure; and encourage investment in innovative technologies.

- Prohibitively expensive research on breakthrough technologies like fusion shall be pursued in international research collaborations initiated, led by and preferably based in the EU or the U.S. Participating countries, including the EU shall continue to adequately finance the International Thermonuclear Experimental Reactor (ITER) to prospectively bring fusion energy to the commercial markets.

Energy Efficiency and Smart Grids

- A new energy efficiency and savings directive with binding targets shall be adopted in the EU, and similar but even more ambitious binding targets for energy efficiency shall be set in the U.S., given the higher per capita level of energy consumption.
- Exchange of information on energy efficiency programs, joint energy efficiency and savings programs on the model of the successful Energy Star Program shall be promoted through the EU-U.S. Energy Council.
- While research investigating future concepts and tools for smart grid operation and planning is underway, closer cooperation to tackle these challenges is needed.
- The U.S.-EU Energy Council's two working groups on energy policy and energy technology RD&D shall be expanded upon to include lessons learned from renewable integration and from electronic vehicle deployment in areas of grid friendly charging, electric vehicles as grid balancers through vehicle to grid technology.
- Closer comparison between areas with substantial renewable energy in the electricity sector, such as the U.S. Pacific Northwest and EU countries Sweden, Denmark or Germany would be opportune to save grid operators from reinventing the wheel.

- Broadening the U.S. Smart Grid Interoperability Panel's scope and including European standard organizations could create a forum to exchange and align ideas.
- Developing a comprehensive approach to formulate smart grid deployment plans would benefit the grid modernization efforts in the United States similar to the EU plans in that area.

Security of Supplies

- Given the interconnectedness of the two economies, it is in the strategic interest of both the EU and the U.S. to ensure the flow of safe, secure and reliable energy supplies. Any disruption might result in serious economic losses for all. A coordinated approach towards third players is of special importance given the energy import dependence of both economies.
- A robust transatlantic energy diplomacy is needed to ensure security of supplies. All available foreign policy tools must be deployed to defend chokepoints and physical infrastructure, eliminate bottlenecks and diversify supplies.
- The U.S. must assist the EU in seeking greater diversity in its suppliers, supply routes and overall energy mix, especially in taking steps to reduce the dependence of European energy markets on single suppliers of oil and natural gas.
- The EU and the U.S. must actively engage with Russia on a number of issues ranging from improving energy efficiency within Russia through European and American investment into Russia's energy sector to ensuring reliable transit through Ukraine on the basis of a rule- and market-based approach.
- The EU and the U.S. must actively consult and collaborate on partnerships with third countries, such as the U.S.-China Energy Policy Dialogue or the Africa-EU Energy Partnership. The EU and the U.S. should coordinate their positions through the EU-U.S. Energy Council before entering into negotiations on energy issues of strategic importance with third players.

- The EU and the U.S. should spearhead efforts to create more stable and predictable global energy markets by reducing volatility, by setting norms and standards and by jointly addressing multilateral rules and governance issues.
- The EU and the US should together engage in a dialogue with major energy producers and consumers around the world in the framework of the newly founded International Energy Forum.

II. Sector-Specific Recommendations

Renewable Energy

- Setting a mandatory minimum renewable energy target in the U.S. on the federal level should be considered.
- The U.S. and EU should maintain the momentum of the current wave of solar energy innovation. This requires the political assuredness to continue costly front-end financing of research and prototype development so that nascent solar energy technologies cross the 'innovation valley of death'. A joint effort, possibly as the marquee initiative of the EU-U.S. Energy Council, could provide both with the necessary mutual confidence.
- Work such as the U.S. National Renewable Laboratory's "Policymaker's Guide to Feed-in Tariff Policy Design" shall be continued at the transatlantic level to enhance systems.
- The U.S. and the EU must renew the stalled efforts to reduce tariffs and non-tariff barriers on environmental goods and services.

Biofuels

- Both the EU and the U.S. have to invest significantly more in the research, development and demonstration of advanced biofuels.
- Support schemes must be feedstock and technology neutral. Government supports have to be predictable and phased out automatically if a technology reaches maturity.

- The EU and the U.S. should enhance their cooperation on the R&D of advanced agricultural technologies and practices.
- As developed nations and technology leaders in the biofuel industry, the U.S. and the EU also share a responsibility for the promotion of technology and best practices of feedstock and biofuel production in the world.
- The EU and the U.S. should coordinate their action on the implementation of sound sustainability criteria for biofuels. The lack of clarity and the uncertainty around sustainability criteria, especially in the EU sends mixed messages to the industry and undermines the credibility of governments' biofuel policies. To avoid such controversies the EU and the U.S. should work together on a coordinated system of sustainability criteria that link financial support to GHG performance and incentivize the economical and sustainable use of natural resources.
- To facilitate international trade, the transatlantic partners should work together in creating international certification schemes for biofuels. These schemes should certify the physical and chemical attributions and the environmental performance of the product, based on life-cycle assessment.
- The EU and the U.S. should continue working together on standardization. The Biodiesel Tripartite Task Force and the Bioethanol Tripartite Task Force created by the United States, the European Union and Brazil to align technical specifications for internationally traded biofuels sets a good example.
- Standardization is also important in order to step over the "blending wall" and enable the proliferation of higher ethanol and biodiesel blends.
- It is worth considering the coordination of targets and mandates applied to biofuels in the EU and the U.S.. Currently the targets of the transatlantic partners are hard to compare and had been established almost exclusively with regard to the development of their respective internal markets.

- The EU and the U.S. should progressively eliminate trade barriers impeding the international trade of feedstock and biofuels. The U.S. Senate's June 2011 decision ending a 54 cent per gallon tariff on imported ethanol is a positive step towards this direction, which should be followed by the EU as well. Eventually the EU will be particularly interested in the free trade of biofuels, since its domestic biomass-producing capacity is limited.
- The transatlantic partners have to make better use of the current framework provided by the Transatlantic Economic Council, the Transatlantic Business Dialogue, the Transatlantic Legislator's Dialogue and the EU-U.S. Energy Council.

Oil and Natural Gas

- The EU and the U.S. should create comprehensive and cohesive policies adopting a stable legislative and regulatory framework that would dramatically enhance efficiency of and encourage investments and development into only those local fossil fuel resources that fulfill a sustainability criteria.
- The EU and the U.S. should work jointly to reduce volatility on the international oil markets, by further enhancing the Joint Oil Data Initiative by the International Energy Forum and stepping up dialogue and cooperation between consuming and producing countries.
- The EU and the U.S. should work towards creating conditions for both gas to gas and gas to non-gas competition.
- The EU should complete the internal market, proceed with market liberalization and unbundling (Third Energy Package), ensure efficient third party access to the infrastructure and continue the steadfast and strict enforcement of EU competition rules in the energy sector.
- The EU should provide matching funds from its budget to develop bottleneck infrastructure and improve energy efficiency, especially in Central Europe and the Baltic and earmark funds for this purpose in the new Connecting Europe Facility from 2014. Building interconnectors and reverse flow

capabilities, completing the North-South Gas Corridor, implementing the BEMIP and NETS interconnection plans, building new LNG terminal in Croatia and in Poland with EU and national assistance are crucial. Regional and sub-regional energy cooperation within the European Union should be strengthened.

- The EU and the U.S. must ensure access to alternative sources and supply routes for Central Europe and Baltic countries with robust energy diplomacy and the EU and the U.S. speaking with one voice; diversification efforts shall be supported by regulatory, diplomatic and financial means.
- The EU and the U.S. should join forces to address resurfacing environmental concerns over natural gas, especially unconventional gas by new regulations if necessary, by transparent industry practice and by increased public and private R&D funding (water issues, fugitive emissions, pipeline leakage, CCS, etc.). The transatlantic partners should develop a safe but sensible regulatory regime on unconventional gas striking a fine balance between the federal and state and between EU and member state levels in the U.S. and Europe respectively.
- The EU, assisted by the U.S., should better manage interdependence with Russia by vigorously using the Early Warning Mechanism, enhancing the energy dialogue, and advancing confidence building measures both in gas and non-gas sectors.
- The EU should ensure that through an unrelenting negotiating position on the Deep and Comprehensive Free Trade Agreement (DCFTA) Ukraine clamps down on corruption in the energy sector and implements the provisions of the Energy Community Treaty in a timely and orderly fashion. Securing and modernizing the Ukrainian transit routes in a trilateral format (Ukraine, the EU and Russia) in close coordination with the U.S. must be a priority.
- A low-profile yet more robust role for NATO can be envisaged primarily in protecting critical energy infrastructure, transit areas and lines, contingency planning for potential disruptions of vital energy supplies.

Nuclear Energy

- Broader global cooperation in nuclear safety is imperative. The International Atomic Energy Agency (IAEA) must be supported and strengthened in its efforts. The U.S. and the EU should work to strengthen international standards and rules in the framework of the IAEA. Each nation has to have a proper regime for regulation and enforcement of nuclear safety and security. Reviews shall be done at the level where the necessary expertise in the specific reactor designs lays, but it is crucially important to undertake systematic peer assessments.

- Emergency preparedness is key and great attention needs to be paid to the eventual loss of onsite power at nuclear plants and the availability of backup power. Prevention is a goal, but rapid response capability must be part of the strategy. Retaining and developing relationships and the depth and breadth of interaction and response that were made during the crisis between all the different stakeholders are important.

- How spent fuel is stored at some reactor sites must be rethought. A reliable emergency plan for coolant loss from spent fuel pools is necessary at every nuclear facility. It is also necessary to rethink the interim storage of spent fuel prior to final reprocessing or permanent storage. It is important to not just improve site-specific plans, but also to establish international and national response capability.

- Effective and uniform liability protection is essential: the Convention on Supplementary Compensation for Nuclear Damage (CSC) should enter into force as soon as possible. Such a uniform global legal regime would compensate victims in the event of a nuclear accident.

- A long-term waste management plan, including guidelines for disposal in a geologic repository should be in place from the beginning of a nuclear power program because reaching agreement on a suitable site for nuclear waste can take a long time.

- Government and industry efforts based on independent, reliable and unbiased research and data shall be increased to improve public acceptance of nuclear energy usage. It is critically important to educate the public about externalities, life-cycle costs and overall impact of all energy resource.
- Loan guarantees and other federal and state level incentives in the U.S. are critical in helping utilities to obtain financing for new construction and mitigating investor concerns about investment recovery, especially given the size of projects relative to utilities' equity base. Industry efforts to deliver plants on time and on budget are also crucial.
- A regional approach in central and eastern Europe is advised to avoid costly public investment into nuclear overcapacities and enhance collective energy security in the region.
- The ongoing dialogue in the framework of the EU-U.S. Energy Council should be reinforced, supporting the development and deployment of safe, low-cost nuclear technology, aiming to develop and promote compatible taxation and incentives policies and focusing on nuclear waste and site issues.
- Recommendations based on the findings of the stress tests in Europe and the NRC review in the U.S. will have to be implemented vigorously and without delay.
- Combating nuclear weapons proliferation remains essential. A renewed international framework for peaceful nuclear cooperation is needed to further minimize the risk of proliferation. The idea of an international fuel bank where nations can commercially access fuel needed for the peaceful use of nuclear energy should continue to be pursued vigorously. Discussions are already underway about the elements of such a regime under the International Framework for Nuclear Energy Cooperation (IFNEC). Instead of building new enrichment and reprocessing facilities, fuel leasing services could serve the needs of both the front and back end of the nuclear fuel-cycle.

Chapter One

Energy in an Era of Unprecedented Uncertainty: International Energy Governance in the Face of Macroeconomic, Geopolitical, and Systemic Challenges

Kirsten Westphal

"The energy world faces unprecedented uncertainty."[1] This is how the International Energy Agency (IEA) started its executive summary grasping the major trends and developments in the global energy relations in autumn 2010—still some months ahead before the Fukushima-Daiichi nuclear catastrophe and the Arab Spring.

This chapter aims to draw a landscape of the uncertain geopolitical and macroeconomic environment of the global energy system. Uncertainty carries risks and opportunities, depending on what actors make out of it. Our existing energy system is internationally connected and intertwined, many segments have a global scope, and the effects of energy production and consumption do not stop at national borders but rather have transnational and global consequences for the climate and the environment.

In writing about "energy in an era of unprecedented uncertainty" one has to be aware of the "knowable" and "unknowable" in future macroeconomic, geopolitical and technological developments. The limited human ability to foresee developments and understand the triggering factors is obvious. It was on display when the macroeconomic, financial and debt crises unfolded in 2008 and 2011 and when the wave of Arab uprisings started in Tunisia and Egypt in 2010/2011. Theory is right in pointing to the fact that human knowledge about the future is limited. We dispose of transparent and

[1] IEA, *World Energy Outlook 2010* (Paris: OECD/IEA, 2010), p. 45.

encompassing information in the case of *known knowns*. As regards the risk dimension, these are well known threats. *Known unknowns* are goings-on where we lack necessary and sufficient information, which cannot be entirely understood and display a certain ambiguity. "*Wild cards*" may fall into this category. *Unknown unknowns* are comparable with *black swans*,[2] being highly improbable and not on the screen of relevant actors. Daase and Kessler enlarge this list by *unknown knowns*: things that are just looked on from the bright sight, ignored or even kept in secrecy.[3] Related risks to these situations are neither communicated nor addressed, making them a blind spot for international (risk) management.

For international energy policy the above is of paramount significance. This chapter aims to shed light on international governance and its (un)preparedness and (in)ability to address the major challenges, to manage the related risks, but also to draw opportunities from a situation in flux. In face of uncertainties, international energy governance should first and foremost be directed to reduce such uncertainties and should provide a stable framework for risk and opportunity management. Stability has always been the key to and the foundation for the energy business, with its long lead times, the life span of once erected infrastructure and the dilemmas of sunk costs. However, this recognition should not lead to perpetuating a given situation, thereby confusing dynamic stability with fossilized stagnation. Uncertainty is also induced by international (non)governance itself. This chapter does not aim to provide an exhaustive overview of existing energy governance mechanisms,[4] but rather points to initiatives with the potential to reduce uncertainties, turn them into opportunities or mitigate related risks. Moreover, it tries to identify certain barriers for the latter. In many respects, governance is still in the process of identifying the risks

---

[2] Nassim Nicholas Taleb, *The Black Swan: The Impact of the Highly Improbable* (New York: Random House, 2007).

[3] Christopher Daase, Oliver Kessler, "Knowns and Unknowns in the War on Terror: Uncertainty and the Political Construction of Danger," *Security Dialogue*, vol. 38, no. 4 (December 2007), pp. 411-434.

[4] For an overview see Dries Lesage, Thijs Van de Graaf, Kirsten Westphal, *Global Energy Governance in a Multipolar World* (Farnham, Burlington: Ashgate, 2010), pp. 51-72.

related to uncertainties and to judge possible damage, the magnitude of the damage and the probability of occurrence.

After the experiences of extreme oil price volatility in the second half of 2008, under the impact of the Deep Water Horizon accident in the Mexican Gulf in April 2010, the Fukushima-Daiichi nuclear catastrophe in March 2011 and the Arab Spring gaining momentum through 2011, risk identification, communication, assessment and management have recently been driving factors for international energy governance. However, new mechanisms and instruments have to be assessed against the related costs of risk management, be it prevention or crisis management. This means no less than a cost-benefit-probability calculation for the relevant actors. Who will take over the costs of enhanced energy (supply) security and of a sustainable energy (r)evolution and to whom is the responsibility attached?

Facing the Triple Energy Challenge—Uncertainties of Political Action

Three global energy challenges demand urgent, simultaneous and multi-level political action: ensuring energy security for every country; minimizing the effects of energy systems on the environment and the climate; and providing access to modern forms of energy to all people.[5] In essence the question is who can take action when, how and to what extent to address effectively the triple challenges of energy security, climate change and equitable access to modern energy sources.

The IEA has been right to point to the urgent need to transform our energy system low-carbon, more sustainable and resource-efficient. Global energy consumption is the major driver of climate change. The energy sector accounts for 70% of greenhouse gas (GHG) emissions. Meeting the internationally acknowledged, albeit non-binding, 'Two-Degrees-Goal,' i.e. to limit global warming by 2050 to an increase of two degrees centigrade in average temperature compared to preindustrial levels, would mean that GHG emissions would have to peak around 2020—yet energy consumption is expected to increase 40% by 2030. As the IEA first presented in detail in its

[5] Aleh Cherp, Jessica Jewell, Andreas Goldthau, "Governing Global Energy: Systems, Transitions and Complexity," *Global Policy*, Vol. 2, Issue 1 (January 2011), pp. 75-88.

World Energy Outlook 2008, this would mean limiting the concentration of greenhouse gases in the atmosphere to around 450 parts per million of carbon-dioxide equivalent (ppm CO_2-eq).

Our energy systems are far from being sustainable, either with respect to the environment and climate or with respect to equity in energy access. More than 1.4 billion people have no access to electricity, which is crucial for economic development. Moreover, more than 2.7 billion people still depend on cooking with biomass. However, energy is not an explicit part of the Millennium Development Goals; it has only lately gained more attention on the international agenda, for instance in the run-up to the Rio plus 20 Conference, as an element in the water-food-energy nexus. Sufficient, stable and affordable energy is not only the key to development and as such a basis of civilization, but clean energy is also a precondition for stable climate conditions and as such for humankind's livelihood and the preservation of the environment.

This is even more true and pressing if one looks ahead to 2050. By then, the world will be populated by 9-10 billion people, compared with 7 billion in 2011. The stress on energy systems and the global climate would be unprecedented. In sum, the sheer magnitude of the triple challenges is daunting, because they are "massive, urgent, global and systemic."[6] Cherp et al. make the strong point that the transformation "will affect how the world produces, transmits and consumes energy and will penetrate all societal levels."[7]

The Global Landscape of Oil Production and Related Uncertainties

One of the main particularities of fossil fuels is their uneven global distribution: in the future, the world will have to rely to an ever larger extent on the energy-abundant countries of "strategic ellipsis:" the geographical area stretching from Siberia to the Caspian Basin, the Persian Gulf to the Arabian Peninsula. The region contains 63.5% of global oil reserves, compared to 47% share in overall production in 2009.[8] The

---

[6] Cherp et al, "Governing Global Energy," p. 76.

[7] Ibid.

[8] BP, *Statistical Review of World Energy 2011*, (London: BP, 2011), pp. 6, 8.

future role of the Organization of Petroleum Exporting Countries (OPEC) and the implications for security of supply rank among the geopolitical uncertainties. OPEC countries control more than 76% of global reserves; 63% are located in the Gulf region. The OECD disposes of only 6% of global oil reserves[9] and conventional oil production in the OECD area has leveled off. For the OECD and its multinational oil (and gas) companies, the time of cheap and easily accessible conventional oil is over. Because of restrictive policies in many energy-abundant states, but also because of depletion paths in the OECD world, the international oil companies (IOCs) have to go to areas that are geologically and geographically ever more challenging.

All of these factors together are driving oil (and energy) prices. The escalation of costs is remarkable: in the 1990s it cost between $500 million and $1 billion for oil field development from exploration until the start of production; today it ranges between $5-10 billion.[10]

As a result, "(t)he size of ultimately recoverable resources of both conventional and unconventional oil is a major source of uncertainty for the long-term outlook for world oil production."[11] The "Peak Oil" discussion has contributed to raise public awareness of the fact that hydrocarbons are exhaustive and non-recoverable. Among the known unknowns is the time when oil production has or will reach its plateau. Estimates differ widely: the Association of Peak Oil (ASPO) argues that the peak of oil production will be reached somewhere between 2005 and 2010; multinational companies are much more optimistic. BP estimates that known reserves will be able to cover demand for the next 40 years. Exploring unconventional oil seems more a matter of cost and price, since new explorations, both off- and on-shore in large "unexplored and untapped" areas, such as deep waters, the Arctic, the Caspian Basin and East Siberia, have contributed to increase the global reserve basis. However, as the Deep Water Horizon accident of April 2010 revealed, these are all potentially related to new significant risks.

---

[9] BGR, *Energy Resources 2009. Reserves, Resources, Availability*, Hannover: BGR 2009, p. 37.

[10] Andreas Oldag, "Eine unbequeme Wahrheit", *Süddeutsche Zeitung*, September 12, 2011, p. 20.

[11] IEA, *World Energy Outlook 2010*, p. 48.

Uncertain Oil Price Developments, Cyclical Investment Swings and Resource Nationalism

Oil prices are set in a complex interplay among market fundamentals, market expectations, financial transactions and speculation. Oil prices are the major reference point for investment in oil exploration, production and infrastructure, and they are the major incentive to reduce oil consumption. Moreover, and this is of utmost importance for the argument of the following section, they determine the state income of oil producing states. Last but not least, they are a lead currency for most other raw materials.

Oil price levels and oil price volatility have been a constant source of concern for all relevant actors. The uncertainties around price development are closely intertwined with the issue of appropriate and sufficient investments. Cyclical investment swings (pig-cycle), albeit being well-known phenomena, have proven to be extremely difficult to cope with. This particularity of the raw materials investment cycle is reinforced by the fact that projects in the energy sector do have remarkable lead times until they are fully developed and on stream, e.g. the respective pipeline is being filled. A certain cycle can unfold: price volatility and uncertainty discourage investment, which in turn prompts country governments to under-invest in productive capacity, a behavior that constrains capacity over time. With the increase in demand, oil prices rise, thereby resulting in a tightening of the supply/demand balance. Both sides take action: in the face of high prices, consuming country governments take action to curb oil-demand growth. Oil demand slows with a lag. At the same time investment rebounds in producing countries, boosting capacity with a lag. This together leads to over-capacity and causes prices again to fall.[12] The (vicious) cycle restarts again.

At the beginning of the 2000s, when a steep increase in demand drove oil prices to new record levels, many oil (and gas) producing states took a more assertive stance towards foreign investments in these strategic sectors. Russia and Kazakhstan alarmed energy investors: re-nationalization and wide scale corruption even challenged fundamental rights guaranteed under production-sharing-agreements. The empiri-

---

[12]IEA, *World Energy Outlook 2010*, p. 141.

cal bases unfolded for Friedman's first law of petro- politics, which states that the price of oil and the pace of freedom move in opposite directions.[13] Whereas the 1990s witnessed a phase of significant inroads into producing countries, the pendulum swung back in early 2000s. Re-nationalization of the oil industry, or at least of its core parts, as exemplified with the Yukos case in Russia, significantly changed the business environment for multinational oil companies, which serve as the major instruments for OECD countries to secure timely, stable and affordable supplies. Today it is the National Oil and Gas companies (NOC) that control over 80% of reserves.[14] NOCs are subject to political considerations and serve as a major instrument for ruling elites to stay in power. Resource nationalism results in limited access for IOCs. It might be simplifying, but their advantage is that they are subject to calculable business considerations and engage in research and development of ever more efficient technologies.

This is where another vicious cycle unfolds. One of the major drivers for the Arab Spring has been price explosions for food. It is here where global oil price increases rebound on energy-abundant states due to the fact that oil prices serve as the lead currency for other agricultural/ raw materials. It does not matter that domestic energy prices are highly subsidized in energy-abundant countries for social, political and economic reasons. The resource curse in most cases of energy-abundant states hinders diversification of their economies. The rent-seeking attitude of the regimes binds the resources for perpetuating the system. This spiral of subsidies is resulting in ever-higher state spending, boosting the state budget. This in turn fuels the global oil price that has to re-finance the additional expenditures.

This is part of the story of why the times of cheap oil are over. Do higher prices guarantee that investments are undertaken? As the *World Energy Outlook 2010* states, "estimates of the world's total endowment of economically exploitable fossil fuels and hydroelectric, uranium and renewable energy resources indicate that they are more than sufficient to meet the projected increase in consumption to 2035."[15] The ques-

---

[13]Thomas L. Friedman, The First Law on Petropolitics, *Foreign Policy*, May/ June 2006, pp. 28-36.

[14]BGR, Reserves, p. 37ff.

[15]IEA, *World Energy Outlook 2010*, p. 117.

tion rather is whether the energy resources will find the money, that is will they be developed in a speedy and timely manner to make them fully available when and where they are needed.

International Attempts to Address Oil Price Volatilities and Investment Uncertainties

It is fair to say that international governance has developed remarkably over the last decade in addressing energy price issues. But it also has experienced some backlashes with regard to investment security.

The Joint Oil Data Initiative by the International Energy Forum (IEF) has been built up as a major initiative to step up dialogue and cooperation between consuming and producing countries. The year 2008 was a watershed for cooperation in the IEF because the decrease of oil prices from $147 to $100 a barrel proved to be equally painful for producing and consuming countries. Since then, there have been efforts to improve transparency in both the oil and financial markets, both under the umbrella of the IEF as well as under the roof of the G20.

But of course there is an ambiguity per se in addressing uncertainties over the demand and supply balance. A shift in any particular direction may result in either a sellers' or a buyers' market, but would certainly bring about significant profits and gains for the relevant actors. Insufficient information and imperfect markets are part of the game. In that respect, market actors naturally object to initiatives by state actors to create greater transparency, in particular with regard to collecting detailed data of investments, as discussed under the JODI umbrella. Uncertainties are only problematic for sellers when they become too big and therefore damage the image of the particular good, or when they encourage buyers to search for alternatives.

While we have seen improvements in tackling high and volatile energy prices, international governance to improve investment stability was dealt a severe setback when Russia ended the provisional application of the Energy Charter Treaty (ECT) in July 2009. Without a doubt, the ECT is a child of the 1990s, when consuming countries managed to make significant inroads into producing countries. The

ECT is an outstanding example of international multilateral governance, aiming for a high level of investment security among its then-51 members.[16] The ECT has produced limited overall results in matters of trade and transit, but its investment provisions generally work well. That was precisely the reason why major energy-producing countries such as Norway and the United States have abstained from ratifying the ECT, because they fear losing sovereign rights.

Lessons learnt from the Arab Spring—More Uncertainties Ahead

With the Arab Spring uncertainty has grown about short- to mid- and long-term developments in the Middle East and North Africa (MENA) region; indeed, the movements mark a watershed.[17] The OECD world had long counted on the autocratic regimes of the region because they seemed to provide stability. This proved to be an mistake: it is "stable societies(,) that hold the key to future reliance on MENA hydrocarbons."[18]

The Arab Spring has caused widespread fears about the prospect of oil (and gas) supply disruptions. Libya was a case in point: regime change resulted in major supply disruptions due to attacks, damaged infrastructure from fighting or as a consequence of international sanctions. In the mid- and long-term perspective any government in the region will have a strong interest in exports, as they offer a major source of income, and an instrument to maintain power as well as to pursue social and economic policies.[19] However, uncertainties stem from internal reforms that may go hand in hand with new depletion

---

[16] Andrey Konoployanik, "Energy Security and the Development of International Energy Markets," in B. Barton, C. Redgwell, A. Rønne, D.N. Zillman, ed., *Energy Security: Managing Risk in a Dynamic Legal and Regulatory Environment* (New York: Oxford University Press, 2005), pp. 47-84.

[17] Hakim Darbouche, Bassam Fattouh, *The Implications of the Arab Uprising for Oil and Gas Markets*, MEP 2, The Oxford Institute for Energy Studies, September 2011, p. 1.

[18] John Roberts, The Arab Revolution of 2011, *Energy Economist*, March 3, 2011.

[19] Kirsten Westphal, *Energiesicherheit und -kooperation auf dem Prüfstand*, in Muriel Asseburg (ed.), Proteste, Aufstände und Regimewandel in der arabischen Welt, Akteure, Herausforderungen, Implikationen und Handlungsoptionen, SWP-Studien 2011/S 27, October 2011, pp. 55-57.

strategies, a revisiting of existing Production Sharing Agreements, changes in the managements of National Oil Companies etc. This in turn, may affect the business conditions for IOCs and the access regime. Moreover, political and socio-economic reforms may affect the volumes and direction of exports, and there are strong indications that necessary investment into new sites is on hold, funds are being redirected and domestic price reforms are being reversed. This is a preoccupying trend for oil and gas supply prospects from the region.

Regionally, there is an immediate risk of contagion and changing balances of power. Iran and the Shiite minorities in the Gulf countries are of concern. This is particularly sensitive when it comes to the three countries of the Gulf Cooperation Council: Saudi Arabia, United Arab Emirates and Kuwait. The sword of Damocles for the global oil markets hangs over Saudi Arabia. The kingdom's strategic importance for world oil markets cannot be overstated. It is not just the sheer size of its production and exports, but mainly its spare capacities to increase oil production at short notice. Saudi Arabia produces around 10 million barrels daily, and has another 2.5 million barrel capacity,[20] which gives the kingdom the opportunity to act as a balancing swing producer. There is no significant spare capacity outside the Gulf region. Its reserves are still easily accessible, of super giant or giant size, with low production costs. More than 40% of global oil exports originate in the Gulf Region.

Risk assessment here easily goes beyond everything what we so far have experienced regarding supply disruptions, because of the potential magnitude with which the energy system, the oil price level and the global economy would be affected. In a certain sense, the image of a Saudi Arabia in turmoil is even a "black swan" because few even dare to think about it. The probability of such an occurrence, of course, ranges beyond the scope of this chapter, but the entire world economy and industrialized countries would be highly vulnerable in such a scenario. Petroleum- based road and aviation transportation is the lifeline of global trade. There is almost no short-term price-elasticity in demand.

[20]Own calculations based on EIA, Saudia Arabia and BP, *Statistical Review*, p. 8.

However, this brings us to the physical trade flows of oil and the major chokepoints in the region. One third of all sea-based oil trade is crosses the Strait of Hormuz. The February 2011 revolution in Egypt raised concerns over the SUMED pipeline and the Suez Channel, through which LNG exports from Qatar to Europe cross. The remarkable gap between West Texas Intermediate Oil and Brent Oil prices made it evident: the U.S. has managed to become increasingly independent from the MENA region. Europe has proved to be much more vulnerable, in particular from trade disruptions in North Africa. Yet, the bulk of GCC crude oil exports go to the Asia-Pacific region.[21] What do the physical oil trade flows tell us? In future, North American oil markets may become less interconnected depending on the amount of deep water off-shore exploration in the Mexican Gulf; unconventional (shale) oil production in the U.S. and Canada; and the prospects for offshore production in Brazil. What does this mean in future for U.S. commitments to ensure exports from the Arab Peninsula to world markets? How will China and India behave with regard to supply risks from the region?

International Crisis Mechanisms and Producer-Consumer Cooperation Revisited

When it comes to dealing with the well-known pig-cycles in raw materials and the shift in the supply and demand balances, the year of the 2008 has been a watershed that has brought energy governance further but not far enough: The drop in high peak in prices from about $147 to $100 per barrel (Brent) in less than seven months raised the awareness of the mutual vulnerability consumers and producers have to volatile oil prices. They bring about a high level of uncertainty. In that respect, the Joint Oil Data Initiative of the IEF is something that has to be backed and developed. Transparency is not easy to achieve, but it is an important asset in an appropriate balance of supply and demand. However, this obviously is a sensitive issue in commercial terms, as buyers favor liquid markets and sellers profit from

[21]For Saudi Arabia the figures are as follows: 57% of crude oil exports and 50% of refined product exports go to the Far East (Energy Information Administration, 'Saudi Arabia, Country Analysis Briefs,' January 2011, p. 6.

tighter market situations. This hinges on the question of a right level of oil prices, which is an eternal source of debate amongst the different players in the market.

The approval of the Charter of the IEF in February 2011 was welcomed as a decisive step forward in producer-consumer relations. The test soon followed, as disruptions in Libya drove global oil prices to sensitive levels. The IEA repeatedly called on OPEC to boost supply. OPEC however failed to achieve a respective consensus in its meeting in June 2011, and major dividing lines became visible among its members. Two weeks later the IEA decided to release 60 million barrels of oil from emergency stocks. This step was criticized at as an attempt to calm gasoline prices. In sum, the two sides failed to produce a coordinated approach to the supply disruption of 1.6 barrel a day of light and sweet Libyan oil. Their respective actions revealed rifts in producer-producer and in producer-consumer relations. It will take time for confidence to be restored, since the actions exacerbated price volatility rather than easing the costs of adjustment."[22]

Uncertainties in Conventional Gas Supply: Wild Cards and Geopolitics at Play

To state the main point directly: the resource base of natural gas is abundant when compared with that of oil, and may easily meet projected demand.

The biggest uncertainty for gas supply again relates to the question whether the investment undertaken will be sufficient and in time. For gas, the share of proven gas reserves in the strategic ellipsis of 71.2% is even higher than oil, and almost 60% is located in four countries: Russia, Qatar, Iran and Turkmenistan. The countries of the strategic ellipsis also dominate gas production with 37.5%.[23] In a global perspective, most of the reserves are conventional gas. The OECD share equals to about 10% share of the world total.

The Caspian Basin, with its vast, partially unexplored and untapped resources, is the region where wild cards and black swans could

[22]Darbouche, Fattouh, *The Implications of Arab Uprisings*, op. cit., p. 21.

[23]BP, *Statistical Review*, pp. 20 and 22.

change the future pattern of resource development and exports.[24] In the Caspian Sea Basin and in Central Asia, geopolitics and geoeconomics are at play, for instance with regard to territorial disputes such as the Russia-Georgia conflict in August 2008; the unclear legal status of the Caspian Sea; policy reversals with regard to depletion strategies, and upstream access, export routes and domestic energy use. Political risk in the region is considered high and these perceptions have preempted larger export projects. At the same time, there are wild cards because parts of the region are unexplored and have promising geological formations, namely in Turkmenistan (as approved by Gaffney and Kline), Afghanistan or Tajikistan.

The region is landlocked, which creates a decisive barrier to the development of its vast resources, notably because of the complexities of financing and constructing pipelines across several countries. With the dissolution of the Soviet Union, and most visibly illustrated in Ukrainian-Russian gas disputes of 2006 and 2009, transit became a, if not *the*, security of supply issue for Europe. Transit issues have been prominently addressed in the ECT and its related Transit Protocol.[25] But at the end of the 1990s, the question of transit (both the rules contained in the Treaty and the Transit Protocol as an annex to the Treaty) became a crucial point of contention for Moscow. The Russian gas company Gazprom feared the loss of its strategically important position as the narrow gateway to Central Asian gas. The decisive point here is that Gazprom buys and resells Central Asian gas instead of simply providing transit services. Russia has put Turkmenistan in the position of being a swing gas supplier, exposing the country twice to two sharp collapses in deliveries to Russia because of disputes over prices and volumes (1997–1998 and 2009). But with the commissioning of the Turkmenistan—China pipeline, gas export has started to diversify. Yet the uncertain legal status of the Caspian Sea so far has prevented the building of a Trans-Caspian Pipeline and the emergence of large scale offshore projects in Turkmenistan.

---

[24]IEA, *World Energy Outlook 2010*, p. 523.

[25]Kirsten Westphal, *The Energy Charter Treaty Revisited. The Russian Proposal for an International Energy Convention and the Energy Charter Treaty*, SWP Comments C08, March 2011.

For the EU, which has repeatedly sought to diversify its imports, it seems very likely that gas exports from the Shah Deniz II phase will find their way through a Southern Corridor into Europe. This might not necessarily have been realized through the Nabucco Pipeline, but rather with a smaller project that is being developed first, and with possibilities to be upgraded later.

In any case, however, these long, overland and multi-country projects make the reliability of exports contingent on a long chain of political arrangements, frameworks and circumstances.

With regard to conventional gas, the creation of the Gas Exporting Countries Forum (GECF) raised concerns, given the dominance of a few exporters and the concentration of reserves in four countries. Pipeline-dominated gas exports and the existence of oil-price-indexed contracts had been seen as factors hindering the formation of a cartel. However, since spot market transactions in Europe and global Liquefied Natural Gas (LNG) trade have increased, the maneuvering room expanded for gas exporting countries to steer the volume and direction of exports. In 2008 the Forum adopted a Charter and set up a permanent secretariat in Doha. The gas glut of 2008/2009 provided incentives to coordinate and discuss production and export strategies among the major players. And indeed, Algeria called for coordinated action in face of depression of spot gas prices in the Atlantic Basin. However, given the then-sensitive situation, the initiative was not taken up effectively by Qatar and Russia. As in OPEC, effective cartelization is constrained by diverging (geo)political and commercial interests of the members. Nevertheless, consumer states should keep an eye on these developments, which are a source of concern.

A Regional Story: MENA Gas and an Uncertain Outlook for Future Exports

The natural gas reserves in the MENA region present some 45% of the world's total, and its marketed production amounts to 20% of the world's total output.[26] Despite the fact that the region is seen as an import asset to (southern) EU gas markets and as an opportunity to diversify EU supplies, as well as the outstanding role Qatar in particu-

[26]Darbouche, Fattouh, Implications of the Arab Uprisings, p. 21.

lar plays regarding LNG, the region plays only a modest role in international gas markets.

The Arab Spring may reinforce certain trends, namely underinvestment and rising domestic demand. Major uncertainties in the short term relate to the political situation in Algeria and Qatar, but so far both countries have not been shaken by uprisings. However, most likely the Arab Spring will affect domestic pricing and market regimes, upstream investment terms, and export policies in the region. The MENA countries will experience steeply increasing demand. This is particularly the case for natural gas, which in turn may constrain the region's export capabilities. Most countries of the region, with the exception of Qatar, have faced supply shortages. Gas will be used extensively in the power sector and as a feed stock for energy-intensive industries, while at the same time upstream production costs will increase as "easy gas resources" are depleting. Gas consumption in the Middle East is expected to grow by 3.9% per annum between 2010 and 2030.[27] North African countries will see a quadrupling of their electricity consumption by 2030, given annual growth rates of 4-8%. This is a major source of uncertainty regarding available export volumes. Algeria is a case in point. Incentives to curb demand will be low, as the Arab Spring most likely will delay and set back domestic price reforms for socio-political reasons.

Rising gas demand and antagonistic relations between neighboring countries are a threat to regional stability. Since the Egyptian revolution in February 2011, the country's gas export contracts with Israel and its other neighbors have been constantly questioned, and the pipeline through the Sinai has been blown up several times. This is a source of geopolitical instability in the region, as Israel gets 40% of its imports from Egypt. Israel could be pressed to exploit disputed offshore resources in the Eastern Mediterranean Sea. The Levant Basin witnessed the world's largest deepwater gas discoveries in 2009 and 2010. This may lead to conflict with the Palestinian Authorities and with Lebanon. Moreover, these vast reserves increasingly have been a source of dispute between Cyprus and Turkey.[28]

[27]BP, *BP Energy Outlook 2030* (London: BP 2011), p. 51.

[28]Vlad Popvici, "Europe's new energy frontier," *European Energy Review*, October 27, 2011.

The fact that production in most countries of the region has not kept pace with demand is very telling: the value-added chain of gas is quite different from that of oil, given the challenges for exporting infrastructure, be it pipelines or LNG terminal. At the end of 2011, Qatar still had a moratorium in place for new gas export projects, pending the outcome of a study of the effects on the reservoirs of the country's North Field, being also the world's largest gas field. The uncertain evolution of Iran and Iraq adds further uncertainty regard the timetable and nature of any production increases.[29]

Shale Gas—A Global Game Changer?

Gas markets have undergone a revolution due to the shale gas boom in the U.S. This illustrates best the uncertainties related to new technologies: fracking was a wild card par excellence. Whether this boom can be reproduced in other parts of the world, namely in (eastern) Europe and China is however uncertain.

The steep increase in shale gas in the U.S. resulted in a remarkable drop in LNG imports to the U.S. This LNG was then exported to Europe, helping to increase the liquidity of its gas hubs and sport markets significantly. The gas glut also unfolded because the economic crisis in 2008/2009 resulted in lower demand. Gas prices eroded in continental Europe and consumers were able to buy gas at a much lower price than under long term contracts with their major Norwegian and Russian suppliers.

As a consequence of these latest developments, gas markets are in flux in respect to their markets, different price regimes and relevant actors. This creates uncertainties. It is very likely that over the next decade, the (North) American gas market will further decouple from the Europe-Asian and the Asian-Pacific market. Due to shale gas, North America will see a price cap on shale gas production costs. The question is open, however, whether the U.S. will become a gas exporter.

Whether Europe will still see an increased availability of LNG (redirected from the U.S.) depends on the level of demand increase in the Pacific region. After the Fukushima nuclear catastrophe, gas

[29]IEA, *World Energy Outlook 2010*, p. 189.

demand is projected to rise, also because of Germany's so-called "Energiewende" and the decision to re-approve the nuclear phase-out of 2002, which had been reversed only some months earlier in 2010.

In continental Europe, oil indexation has come under pressure and we are already witnessing a mix of oil-indexed, gas-to-gas or electricity-based pricing with components of forward spot-market-based elements. This evolution of new price formulas is surrounded by severe doubts as regards risk-reward balancing in long-term contracts. The move toward a new regime means also a break with the former risk-sharing mechanism, under which the buyer took the risk of volumes and the seller the risk of prices under a take-or-pay clause. These contracts were seen as the basis for developing the long value chain of gas upstream projects to transport and distribution nets. There is considerable uncertainty whether this move will secure the needed upstream and infrastructure investments, whether the markets will remain liquid, and whether gas will herald an era of lower gas prices compared to oil or at least whether gas will be competitive when compared to coal in the power sector.

Moreover, these developments have put important gas companies in Europe under pressure. The landscape of European gas companies is in flux, not only because of the uncertain future of long- term contracts and pricing mechanisms but also because of EU internal market regulation. The major uncertainties here relate to the question which actors will be acting in which segment of the gas value-added chain. While on the one hand European gas companies have been faced with unbundling, on the other hand the large gas producers outside the jurisdiction of the EU have actively sought to enter transport, storage and end-user market segments. What this means in terms of price security as one element of security supply still remains to be seen.

In the Asia-Pacific region, major uncertainties stem from a competition between Qatari and Australian LNG about to enter the markets. In addition, the Asia-Pacific region is expected to see the steepest increases in demand and a dominance of oil-price indexed prices due to the Fukushima-Daiichi nuclear catastrophe and to demand increases in the Middle East.

Paradigm Shift Ahead? The Demand Side and Future Energy Trends

It is almost a banality to note that major uncertainties relate to the unknown level and the future structure of energy demand. However, the preconditions are very different around the world: the industrialized countries have almost reached their peak in energy consumption, but the emerging countries and the developing countries have to cope with a (sharp) increase in demand. Over the past 20 years energy consumption grew by 45%, and is likely to grow by 39% over the next two decade. This translates into a 1.7% average growth per annum.[30] The big challenge is to decouple energy consumption from growth of GDP and to decrease energy intensity.

After a short recovery from the deepest economic crisis since World War II in 2010, the debt crises in the U.S. and the eurozone have again shaken financial markets and the world economy. Macroeconomic developments are uncertain, as is the demand for energy. Nevertheless, a pattern is emerging: for the OECD world it seems that due to efficiency gains, the levels of demand after the crisis is unlikely to swing back to pre-crisis levels. 93% of the increase in energy demand will be driven by non-OECD countries.[31] What is unclear, however, is what this means in terms of relative loss of geopolitical weight of the OECD consuming countries and of shifting resource flows, and last but not least for the mix of relevant actors in energy production, trade and consumption. How do and will Chinese and Indian NOCs behave on the world's energy and raw material markets and in trade relations? Will this translate not only into competition over scarce resources but also into conflicts?

Moreover, what does rising demand in energy-producing countries in the strategic ellipsis mean? For supply security considerations of the OECD world, domestic demand in the MENA region, the Caspian Basin countries and in Russia is one of the major sources of uncertainty, and carries significant risk for supply constraints. So far, there are no incentives to tackle the demand side under the paradigm of energy efficiency; to the contrary, cheap and highly subsidized energy prices are the rule and part of the power play of ruling elites.

---

[30]BP, *BP Energy Outlook 2030* (London: BP, 2011) p. 17.

[31]IEA, *World Energy Outlook 2010*, p. 47.

The demand side is key to a more sustainable energy future, as energy savings and increases in energy efficiency are the widest available and cheapest source of energy and thus a low hanging fruit, as has been widely acknowledged. But can the world live up to this potential?

Energy efficiency is the low hanging fruit that is still hanging. Even in the EU, with its 20-20-20 goals by 2020—that is, to achieve a 20% reduction in CO_2 emissions, a 20% share in renewables in energy consumption, and a 20% boost in energy efficiency—the goal to increase energy efficiency is the only one that is non-binding and indicative. Tackling the demand side requires significant steering and administrative capacities of which most of countries do not dispose.

There is another side story to tell: if the EU takes its own objectives regarding GHG reductions seriously and sticks to its commitment to a 2 degrees Celsius track, then this will have a significant impact on the future demand of fossil fuels, which will have to decrease significantly. Uncertainties as well as major opportunities lie in the development of technologies such as CCTS or new fuels in the transport sector. Yet demand issues are highly intertwined with climate security and with supply security. Consuming countries that send the wrong signals with regard to future demand might find themselves in a desperate security of supply situation if they are unable to realize their own plans and estimates.

China, for instance, will have a huge impact on all hydrocarbon markets. Today it accounts for 47% of global coal consumption, and this is expected to increase to 53% by 2030.[32] By 2030 the country's consumption will reach 17.5 million barrels per day, overtaking the U.S.[33] Moreover, gas demand is growing rapidly in China, and by 2030 the country's consumption will be comparable to that of the European Union today. That is remarkable since the share of gas is comparably low in China's energy mix (4% in 2010).[34] China has undertaken steps to reduce its emissions, to a large extent in reaction to local environmental considerations. For gas, major issues are new technologies, developments in the transport sector towards CNG and the price of carbon in the market. Depending on these factors, gas will

[32]BP, *Outlook 2030*, p. 61.

[33]Ibid., p. 33.

[34]Ibid, p. 49.

either remain a "fuel of consequence" or become "a fuel of choice." The question is still open whether natural gas will be only a bridging fuel to a low-carbon energy system, or become a destination fuel, in terms of being the cleanest, vastly available and efficient fossil fuel. BP expects that gas will be the fastest growing fossil fuel in power generation and sees its share in the global electricity growth increase slightly from 20.5% in 2010 to 22% in 2030.[35] The answer to the question is likely to differ between regions and countries.

International governance has only recently focused on the demand side. The most prominent examples are the G8 plus G5 International Partnership on Energy Efficiency Cooperation (IPEEC) and the Sustainable Building Network platforms at the IEA.[36]

Renewables in the Energy Mix: Fundamentals in Flux—Dilemmas at Play

As the IEA outlines,[37] global investment in renewable energy assets grew seven-fold over the period 2004-2008, reaching $126 billion. However, the global financial crisis has hit the renewables sector, which is also likely to be buffeted by the unfolding debt crisis in the EU. Investment in renewables fell 9% and investment in biofuels fell sharply by 60% between 2008 and 2009. The biofuels industry was directly affected by the fall in oil prices and the lower overall demand for oil. Global investment in electricity projects remained stable at around $108 billion between 2008 and 2009.

Major uncertainties relate to the share and structure of non-large-hydro renewables in the overall energy mix. The predominant fuel mix changes relatively slowly, but gas and non-hydrocarbons are projected to gain share at the expense of coal and oil by 2030.[38] The potential for renewables varies greatly however, depending on cli-

---

[35]Ibid., 53.

[36]Dries Lesage, Thijs Van de Graaf, Kirsten Westphal, "G8+5 collaboration on energy efficiency and IPEEC. Shortcut to a sustainable future?" *Energy Policy*, December 2009, online version.

[37]IEA, *World Energy Outlook 2010*, p. 283

[38]BP, *Outlook 2030*, p. 17.

matic, geographical, geological or weather conditions. In order to fully exploit their potential, the world should consider locational advantage and harvest renewable energy under the best possible conditions. This will not come, however, without political, financial and economic costs. It will mean changing the fundamentals of the energy system.

The power sector will be key for enlarging the share of renewables. An increase of end-use electricity consumption will be part of the strategy to curb GHG emissions. Thus far, however, major uncertainties relate to the grid and the interplay between transmission and distribution levels, combining centralized and decentralized power generation. Another uncertainty is related to the development of storage technologies. The shift toward electricity use may help for the 450ppm trajectory, but brings about new challenges for a resource-efficient energy system, as other raw materials such as rare earths are needed for batteries. Here again the question of new technologies is a wild card.

What are the uncertainties ahead? Whether the globe will embark on a more renewable non-hydrocarbon-based energy system, and to what extent it might do so, is unclear. Political and regulatory developments remain particularly uncertain. Moreover, inter-regional and transnational cooperation is a key condition for developing green electricity markets, as reflected in the Seatec concept of large off-shore windparks in the North Sea or envisioned in the Desertec concept. The idea of an energy partnership between the EU and the MENA region in combining wind, photovoltaics, and concentrating solar power (CSP) to the benefit of the MENA countries and to meet the green electricity demand in the EU illustrates the challenges ahead: to break with hydrocarbon path dependencies, to develop and deploy a new technology, make it bankable and commercially viable to achieve the necessary market introduction and penetration. CSP is a case in point that a European technology needs international cooperation in order to achieve effects of scale and cost-efficiency. Whether such North-South partnerships will become real is still an open question due to the lack of supportive economic and political frameworks.

The large-scale use of renewables is not free of geopolitics. It shapes future regions through interconnections. The case of biofuels is closely related to land mass and the water-food nexus. To what extent inter-regional and transnational cooperation can be exploited is still questionable and depends on a level playing field for renewable energy sources compared to fossil and nuclear-based energy sources. Moreover, the development is not free of tradeoffs: any one decision could preempt, constrain or even block alternative solutions. This hints at the systemic character and the complexity of the energy transition challenge. For instance, large photovoltaic installations take up a possible market share for CSP installation, although the latter is able to provide base load.

Without going into detail here, a major source of uncertainty in the development paths of renewables stems from the fact that due to physical, climate and geographical conditions the production patterns of renewables will diversify and differ widely, tailored to the respective situation on the spot. There is no panacea and no one-size-fits-all approach that may spur one global answer. Consequently, the major players have already embarked on different paths. As regards international governance, this has complicated concerted approaches to open question of know-how and technology transfer. The most prominent case in point is the fact that Brazil has not joined IRENA. Here, completely different approaches to renewable energy and clashes over environmental standards are at play.

Coping with Uncertainty and the Limits of International Energy Governance

In sum, in the short and mid-term perspective it is the economic environment that causes the major uncertainties. This study has also tried to shed light on the enormous and related uncertainty about the future price developments, the size of energy resources and the costs to develop them. Moreover, we have seen to what extent new technologies can really be game-changers. But the *World Energy Outlook* is right to point to government policies as "the biggest source of uncertainty to 2035."[39] Whether and which governments will act, and "how, when and how vigorously is far from clear."[40] This is all the more true for international governance.

---

[39]IEA, *World Energy Outlook 2010*, p. 60.

[40]Ibid.

The existing institutional framework is not "fit for 2050" because it reflects the old conventional energy system and there are few loci to address the triple challenge in a really integrated manner. So far, collective action hardly goes beyond paying lip service. It is not so much technological and industrial uncertainties that surround the much needed "industrial energy revolution," but rather "the measures needed to bring this about, the way in which they are to be implemented and their timing are often unclear."[41]

International governance can make a difference. The story behind the G-20's initiative to phase out inefficient fossil fuels, which was taken at the Pittburgh Summit in September 2009,[42] is telling and highlights the related political and regulative uncertainties. Price distortion of fossil and nuclear fuels through subsidies is a decisive obstacle to more efficient energy use, expansion of renewables and effective action on climate protection. If fossil-fuel subsidies were phased out, growth in energy demand would be cut by 4.1%, oil demand would fall by 3.7 million barrels a day and CO_2 emissions would be cut by 1.7 gigatonnes.[43] It is pretty much self-evident that a concerted international effort to reduce subsidies is needed. Energy subsidies interfere massively with national and international energy markets and distort market and price structures. They prevent fair competition between individual fuels and create a cost spiral where renewable alternatives also need greater subsidies. Subsidies on fossil fuels are the root of the evil with regard to price distortions. Their abolishment offers a means to enhance energy efficiency, expand renewables and decarbonise the energy system. Pruning back subsidies can also compensate in a small way for the lack of a market for CO_2. Although this route cannot end the externalization of costs, it would stop the overt promotion of consumption and production of fossil fuels and introduce competitive market conditions. Unless the G20 initiative is

[41]Ibid.

[42]In more detail: Tobias Belschner, Kirsten Westphal, *The G20 and Inefficient Energy Subsidies. Grasping the Cause of Price Distortions by the Roots?*, SWP Comments 2011/C 22, September 2011.

[43]Euractiv, "*IEA top economist calls for bonfire of the fossil fuel subsidies*" http://www.euractiv.com/specialreport-solarpower/iea-top-economist-calls-bonfire-fossil-fuel-subsidies-news-508497?utm_source=EurActiv+Newsletter&utm_campaign=52f296ba43-my_google_analytics_key&utm_medium=email, accessed on October 26, 2011.

implemented consistently and globally, it will be difficult and expensive to tackle the two big challenges in the energy sector: transforming the energy system and eliminating energy poverty. The latest figures for autumn 2011 are discouraging, however: according to the IEA, the equivalent of $409 billion in fossil fuels subsidies are in place, representing a $110 billion increase from the 2009 levels.[44]

The major geopolitical uncertainties relate to fact that the energy system is largely interconnected through price development, but may become less global in the future. The U.S. and the North American market have undertaken significant steps to become largely self-suppliers. This may result geopolitically in a more isolationist attitude towards issues of supply security, in particular in regard to securing oil and LNG flows through major chokepoints. Will Europe and China step in? For Europe, the major challenges to be faced are to develop further the strategic partnership with Russia to the energy relationship with China. Both regions are looking to the same resources in the Caspian region and in Central Asia. It is also an open question how China, as a major consumer of oil and gas flows from the Middle East, will behave as a power invested in ensuring stability and free and open access to supplies. So far, consumer countries' cooperation is limited by the institutional landscape of international governance. Neither China nor India are members of the IEA. However, it is common sense that the consuming countries will be the ones to make the difference and where we can expect action to be taken in regard to climate and energy security.

At the same time, the energy world will become much more differentiated with regard to energy production, consumption paths and the landscape of actors. Major uncertainty relates to the different development paths and their effective implementation. Depending on the locational prerequisites needed for renewable energies, storage facilities etc., the energy production and consumption map will be become much more diversified, being tailored to the circumstances on the spot. This brings about many uncertainties in respect to the development of costs and winning technologies.

---

[44]Ibid.

The landscape of international governance is highly fragmented and rather weak when it comes to addressing the necessary energy transition and at the same time stabilizing the conventional energy (security) of supply without perpetuating it. To put it bluntly: our institutional energy landscape is not 'fit for 2050.' Given the complex nature of relations among actors, the interdependence of energy systems, and interconnections among the energy sectors, as well as the interplay along the whole value-added energy chain, governing or even steering the energy transition is a Herculean task: it demands long-term commitment, shared international energy goals, enormous financial and political resources and integrated energy and climate policies across the different national, regional and global "scales of energy governance, the demand and supply side of energy systems and energy technologies."[45]

There is no single venue to discuss energy issues; instead there is a patchwork of various dispersed, segmented and sometimes competing institutions and organizations. What is equally important is that in many cases the mandate for such groupings was developed decades ago under different circumstances. In many aspects they also no longer reflect existing global energy relations, as traditional roles and boundaries between producers, consumers and transit countries have been blurred. Nor do they mirror the increasing role of non-OECD countries. To make the point very clear, what is missing is not an overarching World Energy Organization. What is needed is a concerted and integrated approach to the triple energy challenge, which as a side product could bring about more coherence in the institutional setting.

It can be reasonably argued that the G-8 has had a good record "as an agenda setter and as a forum for deliberation on energy and climate issues"[46] since the milestone summit in Gleneagles in 2005. Whether the G-20 is able and willing to pursue such an integrated approach is more than open. It would be promising because established and emerging powers meet on an equal footing, but so far the G-20 has

---

[45]Cherp et al, *Governing Global Energy*, p. 75.

[46]Thijs Van De Graaf, Kirsten Westphal, "The G8 and G20 as Global Steering Committees for Energy: Opportunities and Constraints," *Global Policy*, vol. 2, Special Issue on Global Energy Governance ed. by Ann Florini, Navroz K. Dubash (September 2011), pp. 19-30, 28.

taken a more narrow approach to energy. In that respect, it is also telling that sustainable development, with its guiding theme of a green economy, which is supposed to be a major theme at the UNCSD in Rio de Janeiro (Rio plus 20), and has been promoted as an answer to the financial and economic crisis by the OECD, has so far not been taken up seriously by the G-20.[47] As long as the UN Framework Convention on Climate Change (UNFCCC) does not achieve a breakthrough in putting a price on GHG emissions and establishing a global climate regime, and as long as the Post-Kyoto Process remains open, the major elements to spur an energy transition are lacking.

[47]Nils Simon, Susanne Dröge, "Green Economy: Vision mit begrenzter Reichweite," *SWP-Aktuell*, 19 (March 2011).

Chapter Two

Climate Change and the Future of Clean Energy: Toward Transatlantic Convergence

Mark Olsthoorn

If we map how the positions of the EU and U.S. in the arena of climate change and clean energy have changed over the past twenty years (Figure 1), it is clear at first sight that they have not been on converging trajectories.

Figure 1. EU and U.S. positions in the climate and clean energy policy arena, based on author's expert opinion. Framework by Van Soest (2011).[1]

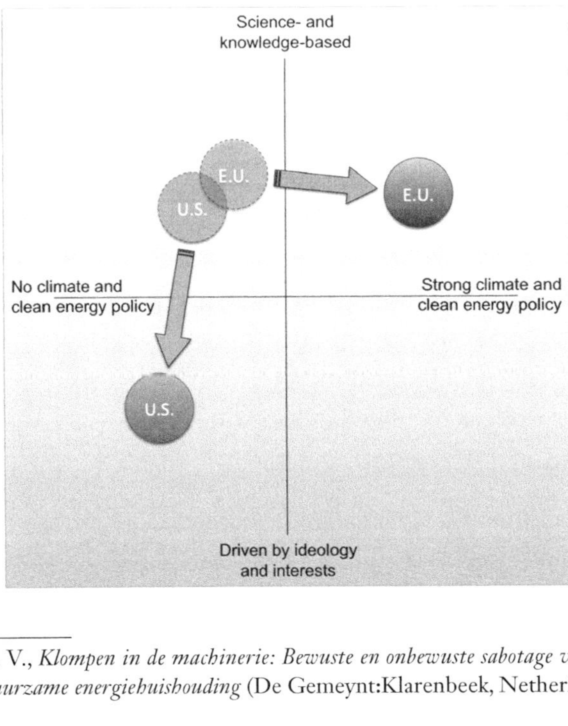

[1] Soest, J. P. V., *Klompen in de machinerie: Bewuste en onbewuste sabotage van de transitie naar een duurzame energiehuishouding* (De Gemeynt:Klarenbeek, Netherlands, 2011).

The trajectories have not progressed along straight lines. While the U.S. in its ambitions has been lagging behind Europe since international climate talks started in the late 1980s, the divergence accelerated during President Clinton's second term, when Republicans took back control in the U.S. Congress, and became official during the Bush years. At first, it was not the idea of man-made climate change that was contentious; the debate focused on what was the best approach. However, over time, a cocktail of free-market ideology mixed with a refined and well-funded 'doubt-machine' of think tanks and media outlets and aggressive lobbying succeeded in making climate change seem like an unproven theory and clean energy like an ineffective pet policy of a socialist government and a waste of hard-earned tax dollars. Europe and the U.S. had become "estranged partners," diverging not only on climate policy, but on a range of issues including defense, biotechnology and international criminal law.[2] When Obama was elected president in 2008, a sigh of relief could be heard in Europe. An era of new rapprochement would soon begin and it was considered a real option that the U.S. would soon align itself with the EU and commit to the UN climate negotiations. In the summer of 2009, the American Clean Energy and Security Act (a.k.a. Waxman-Markey Bill) made it through the House of Representatives. This bill would have required 17% emissions reductions by 2020 and 83% by 2050 compared to 2005; it would have created a nationwide emissions trading system similar to Europe's Emissions Trading Scheme; and included a renewable electricity standard requiring utilities to meet 20% of their demand through renewable energy sources.[3] However, a matching bill failed to pass in the Senate. When Republicans regained control over the House after the 2010 mid-term elections, the odds for new climate legislation being drafted became practically nil, despite an administration that wants to make clean energy a key component of its economic policy.

The doubt-machine has been winning, assisted by persisting high unemployment rates. Polls show that among Americans concern over

[2] Mathews, J. T. (2001). "Estranged Partners," *Foreign Policy*, No. 127 (Nov/Dec 2001), pp. 48-53.

[3] Mehta, S., & Ramig, C. (2010), *Summary of H.R. 2454: The American Clean Energy and Security Act of 2009 (AKA the Waxman-Markey Bill)*. Center for Integrative Environmental Research. College Park, Maryland.

climate change has decreased over the past few years and that fewer people now think that scientists agree that the earth is warming because of human activity.[4] The issue has become highly polarized, with widely different beliefs between Republicans and Democrats.[5]

In the meantime in Europe, worry about global warming remains strong. According to the 2011 Eurobarometer poll,[6] 89% consider climate change a serious problem and 78% agree that fighting climate change and improving energy efficiency represent effective means to economic recovery and jobs growth.

The EU has remained a strong advocate for global climate action and clean energy development and moved forward to devise and implement clean energy policies. The bloc is on its way to overachieving its Kyoto target and it enacted a comprehensive climate and energy package in 2009, setting the famous 20/20/20 by 2020 target: 20% reduction of greenhouse gas emissions compared to 1990; 20% of EU energy consumption to come from renewable sources; and 20% energy savings through energy efficiency compared to projected use. At the core of the climate and energy package are four pieces of legislation.[7] The first is the continuation and strengthening of the Emissions Trading System (ETS), the EU-wide cap-and-trade system. Second are binding national emissions targets for non-ETS sectors that reflect relative member state wealth. Third are binding national targets for the share of renewable energy. Fourth is a legal framework for the development and safe use of carbon capture and storage (CCS). A separate energy efficiency action plan creates incentives toward the energy efficiency target.

The EU Council offered to increase the Union's emissions reduction target to 30% if other major emitters from both the developed and developing worlds were to commit to do their fair share under a

---

[4] Jones, J. M., "In U .S., Concerns About Global Warming Stable at Lower Levels," Gallup.com, March 14, 2011.

[5] "Wide Partisan Divide Over Global Warming," Pew Research Center, April 2008/October 27, 2010.

[6] DG COMM. (2011). *Special Eurobarometer 372: Climate change*. (Brussels, Belgium: European Commission Directorate-General for Communication, October, 2011).

[7] The EU Climate and Energy Package. (2010). http://ec.europa.eu/clima/policies/package/index_en.htm, accessed on October 25, 2011.

global climate agreement. While negotiations under the United Nations Framework Convention on Climate Change (UNFCCC) are ongoing, the economic downturn and a conservative political wind have eroded political support for this higher goal[8]. Hence, despite some divergence, there is an attracting force across the Atlantic when it comes to climate and clean energy policy. Connected markets do not like divergence.

In general, energy policy has three goals: economic competitiveness, security of supply and low environmental impact. The EU and the U.S. hold different views on the most urgent challenges to energy policy, affecting the priority order of their energy policy goals. Besides, the two unions stand in different traditions as to the role of government and markets in dealing with these challenges and have different lifestyles, industries and resources endowments, which also influence the priority order and magnitudes of the policy challenges.

The energy-intensive lifestyle of the average American makes him/her more susceptible than Europeans to changes in the energy system. Low taxes on energy and higher consumption amplify the impact of price fluctuations both in relative in absolute terms compared to most European countries. Securing a reliable supply of cheap energy to keep the economy going and keep prices down is therefore a top priority in American energy politics. Oil is the Achilles heel, because America's economy is highly dependent on road and air transportation, which rely for 99% on oil, and oil is the fossil energy type upon which the U.S. is most dependent for imports.[9] Despite being the third largest oil producer in the world, about half of the petroleum consumed in the U.S. is imported.[10] Natural gas and coal, on the other hand, are almost completely sourced domestically.[11] The fossil fuel industry in the U.S. puts huge economic weight on the table and together with the energy-intensive, low-energy-cost manufacturing industries and lifestyles make for a different and more challenging

---

[8] Lovell, J., "Climate Change Tumbles Down Europe's Political Agenda as Economic Worries Take the Stage," *New York Times*, October 13, 2011.

[9] Energy Information Agency, *Annual Energy Review 2010. Energy*. Washington, D.C., 2011.

[10] Ibid.

[11] Ibid.

cost-benefit trade off when it comes to transitioning to a clean energy system.

Also within the United States, differences in support for clean energy and climate change mitigation play out along lines of differences in current economic structures. States that rely heavily on revenues from coal, oil and gas industries or have built their competitive advantage around cheap base-load power are more likely to resist climate and clean energy policies than energy-consuming states, with little fossil energy resources, high population density and an economy based on services. The opposition may be confirmed by partisan politics, but more likely stems from daily worries of Americans who see their mining, manufacturing and agricultural jobs disappear or at risk and for whom exploration and production of unconventional oil and gas offer hope. Such divides are not alien to European countries, however, in general, European industries and consumers are already used to higher energy prices and the potential of Europe's unconventional fossil fuel reserves offer little hope for a long-term low-cost domestic fuel supply. Therefore it is quite remarkable that within Europe, a heavily industrialized country like Germany, and within Germany industry stronghold North Rhine-Westphalia, are leading the charge and show that even regions whose wealth is rooted in coal, heavy industries and low power prices can gain a competitive edge from climate protection and clean energy development. With effective, bold policies, Germany has leveraged its strength as a manufacturing economy to build a new, clean energy technology sector that is now challenging its car industry as being the largest in terms of jobs.[12]

To which extent 'American exceptionalism'—the belief that America is in a special position in the world and is better than any other country—is a factor when it comes to the persistent state of denial of right-wing America is hard to say. Nevertheless, America has always been a global leader and has shown no intention to accept anything less, given the frequency it is called "the greatest nation on earth" in political rhetoric. It knows what is good for itself and for the world, regarding its free-market based society as the superior model. It does not seem unreasonable to think that such a deep-rooted paradigm

---

[12]Hill, P., Kötter, T., Siegl, E., & Parker, L., eds., *Sharing Solutions: Transatlantic Cooperation for a Low-Carbon Economy* (Washington, D.C.: Heinrich Böll Foundation, 2011).

represents a barrier to accepting that the Europeans are right, that America has missed its chance to lead and should now follow others down the low-carbon road, with massive government intervention in the market place and a restriction of what is considered the American way of life.

In sum, the divergence of the transatlantic partners in the energy and climate arena is obvious, but attracting forces are at work between the poles. America has been holding back the European Union to move to more ambitious climate targets that are more in sync with scientific developments. The widening gap extends beyond the climate and clean energy agenda. A weak transatlantic bond compromises the ability of either the EU or the U.S. to achieve its international aims and reduces the likelihood that other global challenges can be met as well.[13]

We have looked at the differences; we will now see if we can find common ground.

Little Synergy Among Energy Policy Goals

It is often heard that it matters little which of the energy policy concerns one cared most about—climate, energy dependence, or scarcity of resources—we would all find common ground in the solution: a transition away from fossil to renewable energy sources. However, that story is not working—the energy policy goals turn out not to be different sides of the same coin after all.[14]

Fossil fuel resources are finite, so we can be sure that if we continue using them, at some point they will become scarce and production can no longer keep up with demand, resulting in rising energy prices. However, we now know that a true resource crisis is far away. The issue of finite resources may be an urgent one locally, but on a global scale, technically recoverable fossil fuel reserves are abundant. Coal's *proven* reserves alone are sufficient for over a century;[15] with the recent 'discovery' of large reserves of shale gas and other unconventional sources all over the world, natural gas recoverable resources are

---

[13]Mathews, J. T., op. cit.

[14]Soest, J. P. van., op. cit.

[15]*BP Statistical Review of World Energy*, June 2011.

estimated to be more than 200 years of current consumption rates;[16] crude oil may be the least abundant with 46 years of proved reserves,[17] but the industry is venturing into unconventional territories such as shale oil, ultra-deep offshore drilling and above all the Canadian oil sands, which represent many more decades' worth of fossil oil supply. Besides, coal resources and unconventional oil and gas are much more distributed across the globe than conventional oil and gas, reducing the need to transition away from fossil fuels to reduce dependence on sources concentrated in a few areas Europeans and Americans do not like to do business with. Furthermore, there is little doubt that the melting of Arctic sea ice opens up the Northern Ice Sea for exploration and production of significant new reservoirs of oil and natural gas. If one or the other of the fossil fuels were to run out, either form of fossil fuel can be converted into any of the other with known technologies like gas-to-liquids, coal-to-liquids and gasification of coal.

Given these realities, what would a policy for Europe and America look like if they only cared about security of supply? Europe would probably continue to engage in long-term contracts with traditional suppliers, while looking to diversify the supplier list and supply routes, bring eastern Europe into the western European gas network and look for ways to tap into its unconventional resource base. America would do what it is currently inclined to do, especially if the Republican Party got its way: try to rely on domestic production by developing its unconventional oil and gas resources, such as shale oil and gas, deepwater and arctic drilling and connecting to the Canadian tar sands.

When it comes to cost, it is currently the supply and demand of fossil fuels that determine the price we pay for energy. Most low-carbon alternatives still need government support to be competitive. Fossil energy markets are increasingly becoming global and, as we now know, on a global level there is no shortage of energy. Current oil and gas prices offer sufficient incentive to exploit large parts of the huge unconventional resource base. A cost-concerned government would have little incentive to push and pull low-carbon energy sources into the market.

---

[16] *Are We Entering a Golden Age of Gas?* (Paris, France: International Energy Agency, 2011), p. 49.

[17] *BP Statistical Review of World Energy*, June 2011.

However, a government that is most worried about climate change and strives to achieve the internationally agreed goal to avoid dangerous anthropogenic interference with the climate system would have to cap CO_2 emissions at safe levels. The conclusion is that if one is concerned about fossil fuel resources being finite, becoming costly or under the control of rogue regimes that would lead to policies that may have cost-benefits for either of these three concerns, but very little synergy with a climate change agenda. Only energy efficiency would feature on all agendas.

Climate Change is the Most Urgent Energy Policy Issue

In the climate and energy policy arena, both the U.S. and the EU are on a diverging course from scientific developments. The more we learn about the climate, the more we should worry. At the global climate meeting in Cancun in 2010, it was agreed that global average warming should be limited to 2°C (3.6°F) above the pre-industrial average to meet the UNFCCC's mission of avoiding "dangerous and anthropogenic interference with the climate system."[18] This has been the EU's target for some time. However, the impacts of 2°C warming have been revised upwards repeatedly to such an extent that hardly any serious scientist still believes it represents an appropriate "guard rail."[19] A world that is 2°C warmer will put many unique and threatened systems at risk of damage or irreversible loss, from biodiversity hotspots to small island states; it will see large increases in extreme events with substantial consequences for societies and natural systems; despite the impacts being not uniformly distributed, it will see net negative impacts in most regions.[20]

---

[18]see: http://cancun.unfccc.int/cancun-agreements/main-objectives-of-the-agreements/#c33.

[19]e.g.: Smith, J. B., Schneider, S. H., Oppenheimer, M., Yohe, G. W., Hare, W., Mastrandrea, M. D., Patwardhan, A., et al., *Proceedings of the National Academy of Sciences of the United States of America*, 106(11) (2009), 4133-7; Richardson, K., Steffen, W., Schellnhuber, H. J., Alcamo, J., Barker, T., Kammen, D. M., Leemand, R., et al. *Synthesis Report from: Climate Change—Global Risks, Challenges & Decisions*. (Copenhagen, Denmark, 2009).

[20]M.L. Parry, O.F. Canziani, J.P. Palutikof, P.J. van der Linden and C.E. Hanson, eds., *Climate Change 2007: Contribution of Working Group II to the Fourth Assessment Report of the Intergovernmental Panel on Climate Change* (Cambridge: Cambridge University Press, 2007).

What is more, the odds for humanity meeting the 2°C target have become very, very slim.[21] Emissions reduction pathways are being discussed in terms of what is deemed politically and economically feasible, resulting in scenarios that are not in sync with the 2°C target but more closely aligned with at least 3-4°C of warming, bringing high risk of large scale, irreversible changes. A stream of studies have detailed what this means for the economy and society, and with every cycle the picture seems to become more dire.

To bring the climate back in check, to do what is *necessary* instead of what is thought *feasible*, a massive, global effort is required to bring emissions down quickly. Virtually all energy-related greenhouse gas emissions are carbon dioxide (CO_2). CO_2 is a long-lived greenhouse gas that once emitted can remain an active greenhouse gas agent for centuries. Research by Meinshausen et al.[22] and Allen et al.[23] has shown that it is the cumulative emissions that determine the climate impact of CO_2, more so than the emission pathways. Limiting cumulative CO_2 emissions between 2000 and 2050 to 1 trillion tonnes of CO_2 results in a 75% probability of limiting 21st century warming to 2°C above pre-industrial temperatures. Assuming constant emission rates, we would have exhausted the CO_2 budget by 2027 and less than half of currently proven oil, coal and gas reserves can be used (if no carbon capture and storage (CCS) is applied) to meet the target.

In any case, the CO_2 budget is very limited; not exceeding it requires a rapid decline in global carbon emissions. That requires an overhaul of the current energy system, as combustion of fossil fuels accounts for around 94% of CO_2 emissions in both the EU and the U.S.[24] (globally

[21]Anderson, K., & Bows, A., *Philosophical transactions. Series A, Mathematical, physical, and engineering sciences*, 369(1934) (2011), pp. 20-44; New, M., Liverman, D., Schroder, H., & Anderson, K., *Philosophical transactions. Series A, Mathematical, physical, and engineering sciences*, 369(1934), (2011). pp. 6-19.

[22]Meinshausen, M., Meinshausen, N., Hare, W., Raper, S. C. B., Frieler, K., Knutti, R., Frame, D. J., et al., *Nature*, 458(7242) (2009), pp. 1158-1162.

[23]Allen, M. R., Frame, D. J., Huntingford, C., Jones, C. D., Lowe, J. a, Meinshausen, M., & Meinshausen, N. *Nature*, 458(7242) (2009), pp. 1163-1166.

[24]U.S. Greenhouse Gas Inventory (EPA 2011): http://www.epa.gov/climatechange/emissions/downloads11/GHG-Fast-Facts-2009.pdf; European Energy Agency, Climate Data Centre: http://www.eea.europa.eu/themes/climate/dc.

it is about 90%[25]). This will require a much more profound and much quicker change than any of the other energy policy goals.

The fact that cumulative emissions determine the level of warming we will see means that early emission reductions are disproportionally beneficial in delaying the climate change impacts. Studies of the cost-effectiveness of CO_2 abatement options, such as McKinsey & Co.'s famous CO_2 abatement cost curves[26], show that many options exist that can achieve significant CO_2 reductions while saving costs, many of which are measures to improve energy efficiency. By tapping into this potential we can buy time to allow for more innovative low-carbon technologies to be developed, which would reduce abatement costs further down the road, and allow more time for adaptation to climate change impacts that can no longer be avoided.[27] We know very well, as numerous studies have shown (most notably the Stern review[28]), that delaying action will make achieving the 2°C target disproportionally more expensive and more challenging.

The IEA estimates that a portfolio of low-carbon technologies with a cost of up to $175 per ton CO_2 will be necessary to halve global energy-related CO_2 emissions by 2050 compared to 2000 (which would still leave considerable probability of overshooting 2°C), with much larger reductions in western Europe and the U.S.[29] Such a portfolio will include options that can bring scale to the table. In Europe options would include energy efficiency, offshore wind, hydro and biomass and fossil fuels with carbon capture and storage. There is probably an extended role for nuclear power, although limited, as nuclear power is plagued by high financial risks and long lead times.

---

[25]Friedlingstein, P., Houghton, R. a, Marland, G., Hackler, J., Boden, T. a, Conway, T. J., Canadell, J. G., et al. (2010). Update on CO_2 emissions. *Nature Geoscience*, 3(12), 811-812. Nature Publishing Group. doi:10.1038/ngeo1022.

[26]McKinsey & Company, *Impact of the financial crisis on carbon economics—Version 2.1 of the Global Greenhouse Gas Abatement Cost Curve* (2010).

[27]Rhys, J., *Cumulative Carbon Emissions and Climate Change: Has the Economics of Climate Policies Lost Contact with the Physics?* (Oxford, UK: Oxford Institute for Energy Studies, 2011).

[28]Stern, N. H., & Great Britain. *The economics of climate change: The Stern review* (Cambridge, UK: Cambridge University Press, 2007).

[29]*Energy Technology Perspectives 2010—Executive Summary*. (Paris, France: International Energy Agency, 2010).

In the U.S. energy efficiency and fuel switching will be important as well as wind onshore, biomass and solar.

It is clear that preventing "dangerous anthropogenic interference" with the global climate, as internationally agreed in Rio de Janeiro in 1992, represents the most urgent of all energy policy challenges. The extent to which this goal is taken seriously determines the direction and speed of energy policy and investments on both sides of the Atlantic as well as globally.[30]

The U.S. and EU are currently responding to completely different paradigms, leaving little common ground for a transatlantic energy alliance. In 2009 the EU-U.S. Energy Council was established to deepen the transatlantic energy dialogue and address energy security of supply and policies to promote low carbon energy sources, while strengthening the ongoing scientific collaboration on sustainable and clean energy technologies.[31] Unfortunately, it expresses little urgency and can be expected to have little clout in Washington as long as climate change remains taboo in the U.S. Congress.

At the sub-national level, however, meaningful alliances are working across the Atlantic. A number of states have not waited for the federal government to act and have implemented climate and clean energy policies, including renewable portfolio standards, energy efficiency standards, low-carbon fuel standards and cap-and-trade systems.[32] Twenty-nine states and the District of Columbia currently have renewable portfolio standards (RPS), with varying ambitions and sometimes special mandates for specific energy types reflecting their

---

[30]Soest, J.P. van., "Klimaatbeleid gaat vóór energiebeleid," December 17, 2011, http://natuurlijkewereld.blogspot.com/2010/12/klimaatbeleid-gaat-voor-energiebeleid.html.

[31]EU External Action. (2009). The EU-U.S. Energy Council. Available at http://www.eeas.europa.eu/us/sum11_09/docs/energy_en.pdf; Joint Statement Following the U.S.- E.U. Energy Council Ministerial, Lisbon. (November 19, 2010). U.S. Department of State. Washington, DC. Available at: http://www.state.gov/r/pa/prs/ps/2010/11/151185.htm.

[32]Arroyo, Vicki, "Regional Action: A U.S. Perspective," in Phil Hill, Till Kötter, Erik Siegl, Lindsay Parker, eds., *Sharing Solutions: Transatlantic Cooperation for a Low Carbon Economy* (Heinrich Böll Foundation, 2011).

resource base.[33] Twenty-two states have set greenhouse gas reduction goals. In addition, several states have formed regional partnerships to collaborate on greenhouse gas reductions and form cap-and-trade systems. At present, however, only in the northeast did a cap-and-trade system materialize, where power plants in the ten participating states trade emission allowances while collectively working to reduce emissions 10% by 2018.[34] In October 2011, despite fierce opposition from vested fossil fuel interests, the California Air Resources Board adopted cap-and-trade regulation that includes 85% of California's greenhouse gas emissions and will bring the state back to 1990-level emissions by 2020.[35]

Although meaningful and important, it is unthinkable that these bottom-up initiatives will be able to bring about the energy revolution that is needed if there is no strong federal, or even global framework.[36] They should be seen as valuable components but not substitutes for (inter)national climate policy[37]. Besides, concerns over the costs of regional or local programs to affected businesses and their inability to make a dent in global emissions on their own undermine the willingness, or even ability, to run too far ahead of the pack. Under pressure of the economic recession, climate change drops off political agendas; support schemes for low-carbon technologies are being challenged. Some are warning of a "clean energy crisis,"[38] saying that the public subsidies on which clean energy growth have relied are becoming unsustainable in times of fiscal hardship and a setback is around the corner if policymakers fail to implement consistent, long-term policies that focus more on innovation and depend less on subsidies, making them less vulnerable to cutbacks. Such a framework would include binding, science-based intermediate and long-term emissions reduc-

---

[33]Pew Center on Global Climate Change, *Renewable and Alternative Energy Portfolio Standards* (2011). http://www.pewclimate.org/sites/default/modules/usmap/pdf.php?file=5907.

[34]http://www.rggi.org/.

[35]California Air Resources Board adopts key element of state climate plan. *California Air Resources Board*. News Release 11-44 (October 20, 2011); Barringer, F., "California Adopts Limits on Greenhouse Gases," *New York Times*, October 20, 2011.

[36]Van Soest, J.P., *Klompen...*, op. cit.; IEA, Energy Technology Perspectives 2011, op. cit.

[37]Ochs & Spriz, op. cit., 2005.

[38]Victor, D. G., & Yanosek, K. *Foreign Affairs*, *90*(4) (2011), pp. 112-120.

tion goals, public support for R&D and loan guarantees for commercialization projects and market-pull strategies such as a price on carbon and renewable energy standards.

Unfortunately, as long as the U.S. Congress is held hostage to the present state of denial regarding climate change and the inevitability of decarbonization, there is little reason to believe that such a framework will come and little should be expected from a climate and clean energy alliance of national governments on both sides of the Atlantic. However, the tables will turn. And when they do, they can do so very quickly.

Tables can turn anytime soon, as climate impacts become more apparent and the wisdom realized that keeping climate in check is the only sound, economic strategy.

> *"Only a crisis—actual or perceived—produces real change. When that crisis occurs, the actions that are taken depend on the ideas that are lying around. That, I believe, is our basic function: to develop alternatives to existing policies, to keep them alive and available until the politically impossible becomes the politically inevitable."*

That quote by Milton Friedman represents an adequate assessment of the political situation in America and offers appropriate advice to the EU-U.S. transatlantic climate and clean energy agenda. While climate change has disappeared completely from the political agenda in Washington since the midterm elections in 2010, a year that tied 2005 as the warmest on record, the U.S. has suffered from a series of host of extreme events—tornadoes, flooding, drought, and wildfires—that wreaked havoc across the nation. In 2011, a series of spring and summer events broke long-standing records. For instance, by the end of May, unusually high tornado activity had killed over 500 people in the U.S., the highest number since 1953. Rapid melt of an above-average snowpack in the Midwest and extreme rainfall across the Ohio valley combined to cause the Mississippi to reach record-high water levels and flood hundreds of thousands of acres.[39] At the time of writing, the

[39]NOAA, *Spring 2011 U.S. Climate Extremes. National Oceanic and Atmospheric Administration, National Climatic Data Center* (2011). Available online at: http://www.ncdc.noaa.gov/special-reports/2011-spring-extremes/index.php; "Tor-

worst state ("exceptional") of drought has tortured almost the entire state of Texas all summer long.[40] According to the NOAA Climatic Data Center, 2011 economic damage costs-to-date (October 2011) in the U.S. exceed $45 billion, the highest cost-to-date in the U.S. for any year since 1980 when NOAA began tracking "billion-dollar" disasters. Typically, the damage cost-to-date in the U.S. from natural disasters is less than $6 billion, from the usual combination of winter storms, crops losses due to cold weather, springtime flooding and severe weather outbreaks.[41] And not only in the U.S., but all around the globe—from floods in Australia to the horrible drought-caused famine in the Horn of Africa—atypical weather has increasingly been causing humanitarian disasters, killing tens of thousands.

Although such extreme events are not directly attributable to climate change, these patterns are in sync with what scientists say are the consequences of a warming world. And if we ask the question: "would this have happened this way if CO_2 concentrations had remained at their preindustrial level? The appropriate answer is: "almost certainly not."[42] We have entered the era of consequences and, although no one can tell exactly how and how fast, there is no scientific doubt that those will grow worse exponentially if we keep burning fossil fuels unabated. Studies have confirmed over and over again that the economic costs of climate change, even at low levels of warming, are outweighing the cost of stabilizing greenhouse gas concentrations at safe levels. Moreover, climate change does not manifest itself as a gradual change from year to year; it is felt in extremes that come irregularly and unexpectedly, but more frequently. The records of 2010 and 2011 will be broken again, we just do not know when exactly. Even if the U.S. were to save up, an extreme extreme, or a string of extremes, in the U.S. or across the globe, could test the limits of resilience and trig-

---

nado death toll tops 500 and season not over," reuters.com, May 26, 2011. Retrieved from http://www.reuters.com/article/2011/05/26/us-weather-tornadoes-records-idU.S.TRE74P77Z20110526.

[40]U.S. Drought Monitor 2011. http://droughtmonitor.unl.edu/, accessed on October 21, 2011.

[41]NOAA, *Billion Dollar U.S. Weather Disasters*. http://www.ncdc.noaa.gov/ oa/reports/billionz.html, accessed on October 21, 2011.

[42]Hansen, J., "How hot was this summer?" (2010), available at http://www.columbia.edu/~jeh1/mailings/2010/20101001_SummerTemperatures.pdf.

ger a dynamic that leads to a systemic crisis. This is what Paul Gilding refers to as "The Great Disruption."[43] What Gilding says is that the clean energy revolution that is required to prevent such drama is disruptive to many elements of our economy, but if we do not turn the ship around, more likely sooner than later the disruption will be presented by nature. That is when the climate itself is directly speaking to the people, loud and clear, and not solely through 24-hour news networks. And that is why we can be sure that the state of denial in the Republican Party is not sustainable and that what currently is "politically impossible" will become "politically inevitable."

This could happen along three lines: (1) politicians start to see that holding on to the carbon-based growth model is economic suicide; (2) people demand climate action from their elected politicians; or (3) unemployment rates remain high in the U.S. while the EU and China fare well on a model based on a green economy and clean energy and investors start pulling out of carbon-intensive fossil-based stock. The more likely scenario is a combination of these three, reinforcing each other.

This does not mean the EU should wait for this to happen while working on the alternatives. A key transatlantic challenge for the EU remains advocating urgent action, because the earlier the U.S. gets serious about climate change and clean energy, the less disruptive the inevitable process of rapid change will be. When the tables turn in Washington, the EU should be ready. It should know how to move quickly toward global, concerted climate action and a global clean energy revolution that leverages comparative advantages.

We will now move to key elements for a transatlantic energy agenda to decarbonize the energy system.

Elements of a Transatlantic Climate and Clean Energy Agenda

A transatlantic agenda for climate and clean energy has two main objectives:[44]

---

[43]Gilding, P., *The Great Disruption* (Bloomsbury Press, 2011).

[44]Carbon Trust 2009; Victor & Yanosek op. cit., 2011; Levi et al., op. cit., 2010.

1. Turn the main switch to get the train moving: lobbying should continue to get climate change back on the agenda in the U.S. Congress and for the U.S. to get back at the table of the international climate negotiations with a meaningful package of greenhouse gas reduction targets.
2. Align the switches down the road: once denial is left behind, the world needs to move quickly. Clean energy technologies must be developed, improved and get a smooth ride to market. Clean energy markets need to scale up quickly. We know how to do it, but we also know that all boundary conditions are not yet in place. Some switches are still misaligned and the transatlantic partners have a common interest in addressing these barriers.

Although the first objective is key, it is an ongoing effort within the UNFCCC. It is an effort that has many advocates and for which the EU should work with allies across the globe as well as with willing partners within the U.S. We will focus on the second objective, for which the EU and the U.S. are more clearly in the same team, regardless of the U.S. position on binding CO_2 reduction targets.

The brightness of the future of clean energy depends on clean energy technologies becoming cheaper and more efficient. This requires both technological breakthroughs and economies of scale through deployment in the market place.

While China tops the lists of regions where there is most potential for cost-effective investment in CO_2 abatement capital,[45] North America and Europe remain the most important sources of clean-tech innovation.[46] Both have a clear economic interest in facilitating the introduction of innovations into the market and scaling up and integrating global clean energy markets.

An effective clean energy policy employs the right mix of instruments, designed to create cost-effective incentives at each phase of the innovation cycle, from fundamental research in the lab to a technol-

---

[45] McKinsey & Company, *Impact of the financial crisis on carbon economics—Version 2.1 of the Global Greenhouse Gas Abatement Cost Curve* (2010).

[46] Levi, M., Economy, E. C., O'Neil, S., & Segal, A. (2010). *Foreign Affairs*, 89(6), pp. 111-121.

Figure 2. Government support policies need to be appropriately tailored to the stage(s) of a technology's development

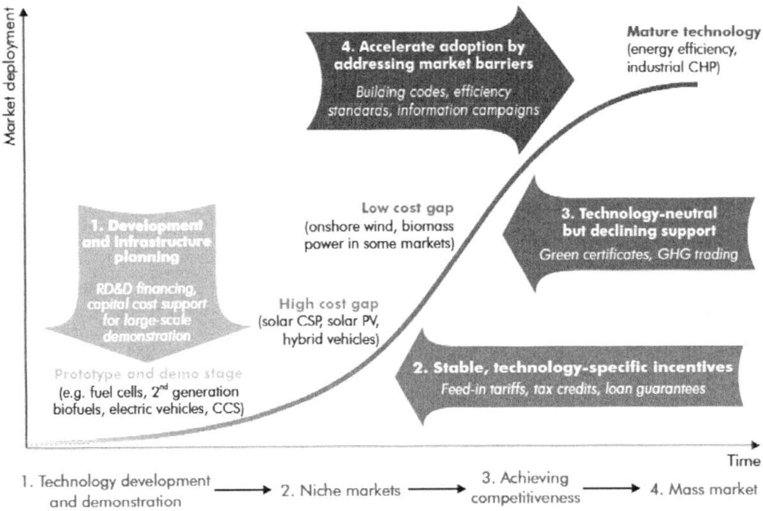

Note: The figure includes generalized technology classifications; in most cases, technologies will fall in more than one category at any given time.
Source: IEA

ogy's competitiveness in the market place. Figure 2, taken from the IEA's 2011 Energy Technologies Perspectives,[47] shows the life-cycle of an innovation from inception to maturity and the various phases along the way.

Each phase has its own market failures that represent hurdles to an innovation's progress. Public policy has a role to play in both innovation and market creation by removing those hurdles. International collaboration can speed things up through standardization, pooling of resources and leveraging each other's strengths.

A Transatlantic Clean Energy agenda should include (but not be limited to) the following set of elements:

- Research, Development and Demonstration: shift funding from deployment to research, development and testing.

[47]IEA, *Energy Technology Perspectives 2010*, op. cit.

- Speeding up domestic and international clean energy markets: provide investment security through stable technology-specific scale-up incentives with a pull character, such as renewable portfolio standards (RPS) and loan guarantees.
- Education: building a skilled workforce.
- Seeking international research collaboration and clean energy market integration.
- Leveling the playing field for energy technologies to allow alternative energy options to compete on their true socio-economic merits.

Shift Public Energy Subsidies Toward Research, Development, and Demonstration (RD&D)

Although many argue that we have the technology to stop the globe from warming more than 2°C, to do so cost-efficiently many of the technologies counted on to play a large role (e.g. carbon capture and storage, solar PV, batteries, smart grid) still require significant RD&D to generate technological breakthroughs and make them more efficient and cost-effective.

RD&D is generally underfunded by the private sector, due to positive externalities ("knowledge spillover," i.e. others get part of the benefit) and highly uncertain return on investment. Government has a key role in funding RD&D in areas upon which the welfare of the people depend, such as energy. However, on both sides of the Atlantic public funding of energy RD&D has been very low during the past two decades and largely directed to conventional energy sources and incremental innovation. In the 2000s public energy RD&D started to rise slowly,[48] and government stimulus packages to mitigate the economic recession of 2008-2009 gave clean energy RD&D a shot in the arm[49], but the IEA estimates that government funding of RD&D needs to be increased 2 to 5 times to achieve its goal of 50% global

[48]Gallagher, K. S., Anadon, L. D., Kempener, R., & Wilson, C., Wiley Interdisciplinary Reviews: Climate Change, 2(3) (2011), pp. 1757-1799.

[49]*Global Gaps in Clean Energy RD&D—Update and Recommendations*. (Paris, France: IEA, 2010).

CO_2 emissions reduction in 2050[50] Governments of both the Major Economies Forum and the IEA have agreed to dramatically increase and coordinate public-sector investments in low-carbon RD&D, with a view to doubling such investments by 2015.[51] Although a good step, that level shows little sense of urgency.

Moreover, the governments should be smart about the projects they invest in or they risk crowding out private RD&D. If public energy RD&D is targeted to more 'risky' and more innovative technologies in the initial stages of development, including prototyping and demonstration projects, the energy sector's innovative capabilities are likely to increase[52] Unfortunately, governments around the world have a strong tendency to support the least risky concepts.[53]

Currently, nearly seven-eighths of clean energy investment worldwide goes to deployment of existing technologies in a market place where most of them cannot compete without government subsidies.[54] Only $ 11 billion of $ 211 invested in renewable energy globally in 2010 focused on technology development—a fraction[55]. For instance, whereas Germany has been very successful in creating stable clean energy investment incentives through its feed-in tariff, by the end of 2010, the real net present subsidy costs of installed solar PV in Germany amounted to € 65.5 billion (2007 Euros, $ 90 billion)[56], which by then was producing only 1.9 percent of the country's electricity consumption.[57] At € 716 (roughly $ 1000) per tonne CO_2 in 2008, solar

---

[50]IEA, *Energy Technology Perspectives 2010*, op. cit.

[51]Ibid.; *Technology Action Plan: Executive Summary*. Major Economies Forum on Energy and Climate (2009).

[52]Garrone, P., & Grilli, L. *Energy Policy*, 38(10) (2010), pp. 5600-5613.

[53]Victor & Yanosek, K., op. cit.

[54]Ibid.; Levi, et. al, op. cit.; *Global Gaps in Clean Energy RD&D*, op. cit.; McCrone, A., Usher, E., Sonntag-O'Brien, V., Moslener, U., Andreas, J. G., & Gruening, C. (Eds.). (2011). *Global Trends in Renewable Energy Investment 2011: Analysis of Trends and Issues in the Financing of Renewable Energy. Management*. United Nations Environment Program and Bloomberg New Energy Finance.

[55]McCrone, et al., op. cit.

[56]Frondel, M., Ritter, N., Schmidt, C. M., & Vance, C. *Energy Policy* 38(8) (2010), pp. 4048-4056.

[57]Federal Ministry for the Environment, Nature Conservation and Nuclear Safety (2011). *Erneuerbare Energien in Zahlen*. Berlin, Germany.

PV was not a very cost-efficient CO_2 abatement option for Germany[58]. While subsidizing solar PV deployment with € 5.0 billion ($ 6.6 billion)[59] in 2010 at seven times the electricity price[60], German government spent only about € 400 million ($ 530 million) on clean energy R&D in 2010 (of which in the order of 30 percent on PV)[61].

Deployment is urgent and needed to bring down costs, but it will be overly expensive if RD&D is insufficient. According to the IEA, the technology areas with the biggest annual RD&D spending gaps are advanced vehicles, CCS, smart grids and industrial energy efficiency.[62] Part of the money needed to boost public RD&D—and lure in private RD&D—can come from substituting more stable pull instruments for politically volatile and inefficient deployment subsidies to create economies of scale for emerging technologies.

Some good examples of transatlantic RD&D initiatives include the partnership of the German Fraunhofer Institute and MIT in the Fraunhofer Center for Sustainable Energy Systems founded in 2008.[63] It is based in Cambridge, Massachusetts, and supported by the Massachusetts government. Another example is the partnership between the U.S. National Renewable Energy Laboratory (NREL) and the Netherlands Energy Research Center (ECN).[64] The two institutions join forces on energy analysis, wind and photovoltaic research.

However, the EU and U.S. should expand their RD&D collaboration initiatives to include emerging economies. Clean energy innova-

---

[58]Frondel et al., op. cit.

[59]EEG Jahresabrechnung 2010. (2011). *Informationsplatform der Deutschen Übertragungsnetzbetreiber.* http://www.eeg-kwk.net/de/EEG_Jahresabrechnungen.htm, accessed on November 2, 2011.

[60]Frondel et al., op. cit.

[61]Federal Ministry for the Environment, Nature Conservation and Nuclear Safety (2011). *Innovation durch Forschung: Jahresbericht 2010 zur Forschungsförderung*. Berlin, Germany.

[62]*Global Gaps in Clean Energy RD&D*, op. cit.

[63]MIT and Fraunhofer announce center for sustainable energy systems, *MIT News*, Cambridge, MA. http://web.mit.edu/newsoffice/2008/fraunhofer-0414.html, accessed on April 14, 2008.

[64]NREL teams with Dutch Energy Research Center. (2009, May 28). *NREL Newsroom*. Retrieved from http://www.nrel.gov/news/press/2009/694.html.

tions, especially large scale technologies such as carbon capture and storage, are cheaper to test there and can be pivotal to emissions reduction in such places as China and India. Some innovation partnerships have already been set up but have remained small.[65]

Speeding up Clean Energy Markets with Smart Push-Pull Policy Mix

If government RD&D policy is effectively generating innovations, they can still easily get trapped in the "valley of death," the phase between demonstration and commercialization, if the appropriate policies are not in place. Dealing with this "commercialization gap" represents the most difficult policy challenge because the costs are greater, it is uncertain how long support is needed and it requires public and private partners working together while maintaining fair market competition.[66] Besides, it is easy to fall into the trap of backing the wrong technologies and becoming dependent on policy instruments for goals they were not intended for (e.g. using instruments intended for commercialization to meet emission targets).

A well thought-out combination of "technology push" and "market pull" instruments is needed to create an efficient clean energy technology innovation and commercialization engine. Each extreme has failed in the past. Governments have never been good at picking the winning technologies of the future, but leaving it all to the free market did not bring forward clean energy innovations as quickly as required. This path through the middle is what the Carbon Trust calls a "technology focused approach." Key elements are the following:[67]

- Priority technologies should be identified through a systematic assessment, taking into account the importance of the technology toward the national energy goals and the comparative advantages.

[65]Victor & Yanosek, op. cit.

[66]Ibid.

[67]Carbon Trust *Focus for success: a new approach to commercialising low carbon technologies* (London, UK: The Carbon Trust, 2009).

- Instruments should be tailored to a priority technology (or technology group) and the development stage a technology is in. The further it has progressed the more the market forces should be leveraged.
- Ensure strong competition between companies within a family of technologies.
- Close and careful monitoring against milestones to minimize waste.

Public policies should generate sufficient investment security for private funders to invest in early application of emerging technologies and implement market pull policies that allow economies of scale to bring down cost. Loan guarantees, feed-in tariffs and performance standards are instruments that can be successful in these phases if designed appropriately.

Feed-in tariffs have become the instrument of choice among EU member states to increase their share of renewables and meet the binding targets set as part of the EU Climate and Energy Package. Twenty of twenty-seven EU member states have implemented them, following Germany's example.[68] In the U.S., renewable portfolio standards are more popular. Twenty-nine states plus the District of Columbia have them, although the set of qualifying technologies and the level of ambition varies widely.[69]

These commercialization instruments should be combined with technology-neutral instruments that level the playing field with incumbent technologies in the energy market, such as outreach and education, standardization and providing equal access to the market and ways of pricing negative externalities, of which CO_2 makes up the largest share by far.[70]

The most (cost-)effective instruments are yet undetermined. However, many nations and states have designed their policies differently.

---

[68]http://www.renewableenergyfocus.com/view/15892/status-of-feedin-tariffs-in-europe-2010/.

[69]Database of State Incentives for Renewables & Efficiency (2011). RPS Policies, September 2011. Available at www.dsireusa.org.

[70]CE Delft. *External Costs and Benefits of Electricity Generation*. Delft, Netherlands, 2010.

Many have difficulty depoliticizing them, but for these programs to be more effective they should be put at arm's length of from day-to-day government operations.[71] The current policies provide valuable data points that should be evaluated in a transatlantic dialogue to determine best practices and promote their dissemination. When the tables turn in the U.S., American and European researchers and policy makers should know which policies work best.

Furthermore, the success of many instruments depends on their continuity and the confidence investors have in them. As clean energy instruments are subject to waves of external pressures—such as resistance from vested interests, budget hawks, occasional failures, ideologies –transatlantic partners should support each other in resisting the temptation to compromise too much on key success factors of the policies at hand.

Skilled Workforce

Although the policy framework is crucial, the heart of the clean technology revolution is science, technology, engineering and construction. If this revolution actually can happen is contingent upon a large enough workforce with the right skills, skills in science, technology, engineering and mathematics (STEM). This is an essential area of transatlantic cooperation. Business leaders and scholars, however, have been warning for a 'skills shortage.'[72]

It is an issue that is said to threaten competitiveness in both the U.S. and EU. Business leaders are calling for government intervention to take measures to promote STEM education and training. If the shortage persists, it could turn into a potential barrier to clean energy market growth. When the main switch finally turns and the switches down the road are aligned, the train may not move because there are too few to get it up to speed.

---

[71]Victor & Yanosek, op. cit.

[72] "Special Report: European Business Summit to Highlight Skills Shortage," Euractiv.com, May 17, 2011; "Skills shortage threatens U.S. manufacturers," *Financial Times*, October 17, 2011.

International Research Collaboration and Clean Energy Market Integration

High on the transatlantic climate and clean energy agenda should be engagement with emerging markets and the promotion of cross-border collaboration on research, development, demonstration and deployment, the free flow of technology and global competition. We know that innovation works best when scientists, entrepreneurs and governments can build on each other's successes, strengths and comparative advantages. However, some pitfalls need to be avoided.

Clean energy is often touted as a way to boost national competitiveness. Countries like Germany explicitly support clean energy technology to establish themselves as leaders in the market of the future. Such a clean energy race can create a virtuous cycle in which competition fosters innovation. However, some[73] warn that with an energy agenda narrowly focused on national competitiveness, there is a risk that the virtuous cycle is overpowered by a vicious cycle of what Levi et al. (2010) call the 'Balkanization' of clean energy markets. That is when markets get fragmented, when among fierce international competition, governments feel compelled to erect barriers and shield off their clean energy sectors for foreign participation or promote their own standards in an attempt to put their own industry at a competitive advantage. This would be a dangerous, lose-lose situation, given the urgency of the task at hand. It stifles international collaboration, slowing down the innovation so badly needed.

Nearly all of future growth in energy demand and, CO_2 emissions and investments in energy infrastructure comes from developing nations, especially the large, fast-growing economies such as the BRICS nations (Brazil, Russia, India, China, South Africa).[74] Given their high growth rates, they must implement clean energy options very fast. Some of them are quickly becoming important developers and manufacturers of these technologies, while the OECD countries are still the engine of innovation. A priority for a transatlantic clean energy partnership is to foster the spread of low-carbon technologies to the developing nations, but also promote the flow of technologies

[73]Levi, et al., op. cit.

[74]IEA, *Energy Technology Perspectives 2010*, op. cit.; Victor & Yanosek, op. cit.

from those areas. This includes reciprocal efforts to open markets for foreign investments and optimal, innovation-effective protection of intellectual property rights in developing countries, especially China.[75]

Leveling the Playing Field

In the energy marketplace, clean energy technologies are competing with conventional sources. One key barrier to clean energy technologies being able to survive without government subsidies and their widespread deployment is that the playing field is skewed in favor of fossil energy. First, fossil fuels receive hundreds of billions in financial support from governments; the current infrastructure has evolved around fossil fuels and centralized power generation and is maladapted to larger shares of distributed or intermittent generation; and external costs are allocated elsewhere and not reflected in the market price, such as defense spending to secure stable oil supplies; health costs associated with air pollution; and most importantly CO_2 emissions of fossil fuels.

Fossil fuels keep receiving hundreds of billions of financial support from governments each year in the form of tax breaks and subsidies.[76] Research by Bloomberg showed that in 2008 fossil fuels received twelve times more government support than renewable energy[77] (on a per unit of energy basis the ratios are different). Fossil fuel subsidies date from earlier times and were intended to promote investments in uncertain exploration and production of domestic resources as part of energy security policy. In developing countries they are intended to promote economic development and alleviate poverty. However, the legitimacy of these subsidies are repeatedly being questioned, as they come with a host of unintended consequences, ranging from discouraging investments in energy infrastructure to disproportionally benefiting the rich. They keep fossil fuels artificially cheap, subsidizing global warming and shifting the profitability threshold for clean energy alternatives in times when the opposite is needed, while often

[75]Levi, et al., op. cit.

[76]*IEA analysis of fossil-fuel subsidies*, International Energy Agency. Paris, France. http://www.worldenergyoutlook.org/files/ff_subsidies_slides.pdf, accessed on October 14, 2011.

failing to meet their intended objectives. As such they result in an economically inefficient allocation of resources and market distortions. There is good reason to believe that most of these subsidies are not needed, given the healthy margins the International Oil Companies (IOCs) keep reporting.[78]

In 2009, both the G20 and APEC[79] agreed to "phase out inefficient fossil fuel subsidies over the mid-term."[80] Although improvements were made in India, China and Russia,[81] according to the IEA fossil fuel subsidies amounted to $409 billion in 2010, up from $312 billion in 2009, and without further reforms are set to reach $660 billion in 2020 (0.7% of global GDP).[82] IEA Chief Economist Fatih Birol says that cutting such subsidies in major non-OECD countries "is the one single policy item" that could help get the world back on track toward no more than 2°C global warming.[83] However, when even in these times when the U.S. government is desperate to reduce its budget deficit and repealing fossil fuel subsidies appear to be a no-regrets option, they still represent a political no-go. The EU is grappling with the issue, too. That is unfortunate, because this is an issue in which the transatlantic partners should be playing as a team and show leadership, starting with forward-looking reform on their own turf.

---

77 "Fossil Fuel Subsidies Are Twelve Times Renewables Support," *Bloomberg*. http://www.bloomberg.com/news/2010-07-29/fossil-fuel-subsidies-are-12-times-support-for-renewables-study-shows.html, acessed on July 29, 2010.

78 e.g. Blery, M., "EOil Companies' Profits Track Gasoline Prices," *Forbes*, October 21, 2011. http://www.forbes.com/sites/sageworks/2011/10/21/991/.

79 Asia-Pacific Economic Cooperation, www.apec.org. 21 member states around the Pacific cooperating on trade and economic issues.

80 Lang, K. *The First Year of the G-20 Commitment on Fossil-Fuel Subsidies: A commentary on lessons learned and the path forward*. Global Subsidies Initiative, International Institute of Sustainable Development. Geneva, Switzerland (2011). http://www.globalsubsidies.org/files/assets/ffs_g20_firstyear.pdf.

81 "IEA warns of ballooning world fossil fuel subsidies." Reuters, October 4, 2011. http://www.reuters.com/article/2011/10/04/us-iea-idU.S.TRE7931 CF20111004.

82 *IEA analysis of fossil-fuel subsidies*, op. cit.

83 IEA top economist calls for bonfire of the fossil fuel subsidies. (2011, October 24). *EurActiv.com*. Retrieved from http://www.euractiv.com/specialreport-solarpower/iea-top-economist-calls-bonfire-fossil-fuel-subsidies-news-508497.

Finally, with or without subsidies, current arrangements still make investments in fossil energy capacity more profitable than investments in clean energy. For instance, the U.S. Department of Defense between 1976 and 2007 spent $7.3 trillion on aircraft carriers in the Persian Gulf with the explicit mission to secure oil shipments.[84] Such costs are not allocated to fossil fuels but end up being spread out over U.S. taxpayers. Moreover, the use of fossil fuels causes a host of unintended negative consequences, from local air pollution to global warming. There are costs associated with these impacts that are not reflected in the market price of energy. These 'external costs' can be large. A study by the U.S. National Academies of Sciences found that burning fossil fuels costs the United States about $120 billion a year in health costs, mostly because of thousands of premature deaths from air pollution,[85] compared to $1 trillion in annual energy spending (2005).[86] However, CO_2 emissions generate the largest external costs by far. For coal-fired power plants CO_2 emissions account for 70-85% of total external costs.[87] Estimates of the costs associated with each ton of CO_2 emitted (the social costs of carbon: SCC) vary widely and range from several tens of dollars per ton CO_2 to $300-900 per ton, rising further over time ($300 per ton CO_2 translates to $3 dollar per gallon of gasoline).[88] All estimates arrive at much higher costs than the $21 used by the U.S. government in its policy evaluations and the €10-15 for which a ton of CO_2 trades in the EU Emissions Trading Scheme.[89]

Full internalization of external cost would dramatically improve the market position of clean energy technology. However there is little hope that this can happen without a long-term global climate frame-

[84]Maass, P., "The Ministry of Oil Defense," *Foreign Policy*, August, 2010. http://www.foreignpolicy.com/articles/2010/08/05/the_ministry_of_oil_defense? page=0,0.

[85]*Hidden Costs of Energy: Unpriced Consequences of Energy Production and Use* (Washington, D.C.: The National Academies Press, 2009).

[86]EIA, *Annual Energy Review 2010*, op. cit.

[87]Bennink, D., Rooijers, F., Croezen, H., Jong, F. de, & Markowska, A., *External Costs and Benefits of Electricity* (Delft, Netherlands, 2010).

[88]Ackerman, F., & Stanton, E. A. (2011). *Climate Risks and Carbon Prices: Revising the Social Cost of Carbon*.

[89]Soest, *Klompen...*, op. cit.

work, but without it the market will keep preferring fossil options more often than our climate goals permit.

Little They Cannot Accomplish

Although the U.S. and EU leadership seem to have diverged on climate and clean energy policy, the EU and the U.S. should keep investing in a strong transatlantic collaboration on clean energy and climate. This chapter provided suggestions for areas where common ground can be found and which collaboration should focus on to bring climate to the center of energy policy and improve the odds for a global clean energy revolution once the tables turn in the U.S. Congress. While pushing for reengagement of the U.S. federal government and building moment through regional and state initiatives, there are many areas where the EU and U.S. find synergy in collaboration and mutual learning at the federal level. These include clean energy research, development and testing, policy evaluation and testing, global clean energy market integration, workforce development and joint leadership on fossil fuel subsidies. Effective collaboration on these issues might very well contribute to convergence in the global climate arena. As Jessica Mathews wrote, "When the United States and Europe see eye to eye, there is little they cannot accomplish. When they do not agree, however, there is little they can achieve."[90]

[90]Mathews, *Foreign Policy*, op. cit.

Chapter Three

The European Carbon Trading System after 2012—Implications to the U.S.

Pál Belényesi

The effects of climate change are undoubtedly real. Much has been said and disputed about how to "environmentally" tackle problems arising from the effects of global warming as well. As it is not the aim of this chapter to discuss the possible environmental implications of these issues and to contribute to those ongoing discussions, it should suffice to show briefly how greenhouse gases contribute to global warming.

The contribution of greenhouse gases to global warming.[1]

Type of gases	Contribution to global warming in %
Dinitrogen-oxide	6
CO_2	55
Freon	24
Methane	15

Environmental regulation could target Pareto efficiency—maximizing net benefits (while minimizing costs)—or maximize cost-efficiency (trying to achieve a given level of environmental protection while pushing down the costs of regulation). There are several other factors that might be taken into account: information sharing, overall effectıveness, equity, monitoring and social effects. Emission trading is only one instrument by which actual benefits might be able to be gained after 2012.

In this chapter I will focus on how a certain approach that was designed to tackle these challenges has been designed in Europe and

[1] Cs. Pánczél, *Pénz- és tőkepiacai eszközök a klímaváltozás mérséklésére*, 2009. http://elib.kkf.hu/edip/D_14877.pdf, accessed on August 23, 2011.

how it will change after 2012, when the first results of the Kyoto undertakings should appear at the global level.

Emissions Trading from an Economic Point of View

Economically effective goals should be defined in the field of climate protection. There is a need to find the method of conditioning that leads to the most expense-efficient achievement. In 2003 Stern observed the global effects of climate change by its probable economic aspects.[2] He pointed out that—economically—the advantages of firm and early actions are more considerable than the likely costs. He underlined that the poorest countries with weak economic structures are more exposed to the negative effects of climate change than others. Similarly, developing countries are definitely at a disadvantage, even though their CO_2 emission is much lower compared to some countries with developed economic structure.[3] Yet, to the contrary, certain countries (e.g. Russia, Canada and the Scandinavian Peninsula, Hungary) could even profit from some degree of global warming, because one of the probable consequences of global warming is that the agricultural output of these countries might increase.[4]

A special program, the so-called "system of distributable emission quotas" (the cap and trade system), assures in general the quantity of the pollution allowed by the authority (or a given international contract), that is to say a system like this is appropriate for the goals of environmental protection. As for the economic side of the system, it is important to emphasize that the related costs are unclear because the price of the quotas is determined by the market, therefore the price of the quotas (and the opportunity cost of the quotas) can change within a wide interval. Practically, the result is that carbon emissions become

---

[2] The Stern Review on the Economics of Climate Change, 2006, accessible here: http://webarchive.nationalarchives.gov.uk/+/http://www.hm-treasury.gov.uk/d/Transport_annex.pdf, accessed on September 10, 2011.

[3] For a commentary on the review by the author, see: N. Stern, "What is the Economics of Climate Change?" *World Economics*, Vol 7, No 3, April-June 2006, pp. 1-10.

[4] Cs. Pánczél, *Pénz- és tőkepiacai eszközök a klímaváltozás mérséklésére*, 2009. http://elib.kkf.hu/edip/D_14877.pdf, accessed on August 23, 2011.

new agents of production for the stakeholders, just like funds, manpower, land and other natural resources.

The emissions trading system established by the European Union (EU ETS) is also a cap-and-trade system, i.e. a trading scheme of fixed total quotas: the quantity of total emissions is ex ante fixed and the stakeholders are allowed to sell or buy emission units within this limit. One unit allows the emission of 1 tonne of CO_2. The word "cap" means the limit value, while "trade" refers to free trading within the limit value.[5]

The European ETS was launched after a series of long international negotiations in 2005 concerning the means for dealing with climate change. This process—as it is not the subject of this study—can be presented briefly as follows:

- After a long conciliation period, in June 1992 the United Nations Conference on Environment and Development (UNCED) was organized in Rio de Janeiro, in which more than 117 countries participated. At the Conference the United Nations Framework Convention on Climate Change—UNFCCC was signed. The Convention came into effect in March 1994.
- In 1998 the UNEP (United Nations Environment Programme—Environmental Protection Programme of the UN) and the WMO (World Meteorological Organisation) established the International Climate Change Partnership (ICCP), the aim of which was to overview and interpret observed data, and to conduct modelling and effect analysis.
- A substantial number of UN member states signed the Kyoto Protocol, in force for the period 2008-2012. The protocol aims at reducing—via national measures—greenhouse gas emissions by 5.2%.[6] The costs of the commitments were attempted to be reduced by including three flexible mecha-

[5] Emissions Trading Scheme (EU ETS), official site of the European Commission, 15 November 2010, http://ec.europa.eu/clima/policies/ets/index_en.htm, accessed on August 23, 2011.

[6] February 16, 2005: The Kyoto Protocol came into effect without the participation of the United States.

nisms, centrally stimulating sustainable development through investments and, where possible, technology transfers. The mechanisms are: the clean development mechanism (CDM), the joint implementation plan (JI) and emissions trading (ET). The latter two are project-based mechanisms.

- The member states of the European Union affirmed the Kyoto Protocol on 31 May 2002.[7] Directive 2003/87/EC set up the EU Emissions Trading System (ETS)[8] that identified the emitters to be regulated.[9] The ETS was finally launched in January 2005. Then, in October 2005 the second European Climate Change Programme (ECCP II) was launched. In 2007 the European Commission presented the Energy and Climate Change Programme and made a proposal to reduce greenhouse gases by 20% compared to the level of 1990. The goal of the two ECCPs was to identify and develop all the essential elements of an EU strategy to implement the Kyoto Protocol.[10] The second ECCP tried to match the Lisbon goals of the EU—more stable economic growth and more jobs—with the exploration of cost-effective options for reducing greenhouse gas emissions.[11]

---

[7] Beliczay, Szabó, *Az éghajlatvédelem gazdasági eszközei—Az emisszió-kereskedelem* (Budapest: Levegő Munkacsoport, 2003).

[8] European Emissions Trading System. Directive 2003/87/EC of the European Parliament and of the Council of 13 October 2003 establishing a scheme for greenhouse gas emission allowance trading within the Community and amending Council Directive 96/61/EC. OJ L 275, 25.10.2003., pp. 32-46.

[9] Directive 2004/101/EC modified Directive 2003/87/EC. (see: Directive 2004/101/EC of the European Parliament and of the Council of 27 October 2004 amending Directive 2003/87/EC establishing a scheme for greenhouse gas emission allowance trading within the Community, in respect of the Kyoto Protocol's project mechanisms, OJ L 338, 13.11.2004, pp. 18-23).

[10] See further: http://ec.europa.eu/clima/policies/eccp/index_en.htm, accessed on September 28, 2011.

[11] See further the European Commission's site for the ECCPs: http://ec.europa.eu/clima/policies/eccp/second_en.htm, accessed on September 28, 2011.

The 1997 Kyoto Protocol

The Kyoto Protocol represented an important step forward in the effort to tackle climate change; it includes binding, quantified objectives for limiting and reducing greenhouse gases. The main focus of the Protocol is to decrease carbon dioxide emissions.[12]

The parties that signed the Protocol committed themselves to comply with the policies and measures defined in the Protocol, including:

- raising energy efficiency concerning certain branches of economy;
- supporting sustainable agriculture;
- introducing measures which aim at reducing emission in the transportation sectors;
- supporting the application of renewable energy sources and environment-friendly technologies.

In total, 38 industrialized countries and countries in transition undertook the reduction of emission by 5.2% for the period of 2008-2012.[13] Among the larger industrialized countries, only the U.S. did not participate.

The Protocol tackles the emissions of the following six gases: carbon dioxide, methane, nitrous oxide, hydro fluorocarbon, per fluorocarbon and sulphur hexafluoride. The denomination of the last three compounds by the Protocol is significant because their atmospheric presence is of especially long duration. The commitments concerning the emission regulation of the involved countries are valid for all these six gases in a way that the emission changes of these gases can be defined with the global warming potential (GWP), the relative measure of how much heat a greenhouse gas traps in the atmosphere.

---

[12]For a more detailed overview, see e.g.: Grubb et al, *The Kyoto Protocol, a Guide and Assessment*. Energy and Environmental Programme, RIIA, 1999. For a more in-depth analysis on the compliance costs, see e.g.: A. S. Manne et R. Richels, *The Kyoto Protocol: A Cost-Effective Strategy for Meeting Environmental Objectives?*, (Stanford: Stanford University, 1998).

[13]For the EU, this number is 8%, with Poland and Hungary being exempt. They have to reduce their emissions only by 6%. Malta and Cyprus are not party to Annex I of the Protocol.

The countries agreed to the emissions limitation and reduction in the percent of base level defined in the Protocol. According to this agreement, for certain countries the commitment means emissions limitation (e.g. Norway, Iceland, Australia), other countries are only committed to retain the emission level (e.g. Russia, Ukraine, New Zealand), while other countries are committed to reduce the emission by 5-8%.

The EU ratified the Kyoto Protocol in May 2002 and the member states of the European Union defined 8% as the value of emissions reductions, but then they set different emissions regulation levels for themselves. They agreed that they must achieve the undertaken commitments together, so slight deviations in the national quotas are accepted.[14]

It remains to be seen, however, how the targets are to be implemented and how it is possible to enforce the undertakings by the parties effectively.

The Story of EU Emissions Trading Regulations

Regulations on the subject of the European CO_2 trading may be divided into three phases.[15] The first steps were taken before 2000. These actions aimed at preventing and reducing climate change at the global level. In December 1997 governments signed the Kyoto Protocol, which legally obliged them to reduce the emission of greenhouse gases. In 1998, the EU differentiated the 8% target between the different member states in the so-called EU Burden-Sharing Agreement.[16] The idea behind this agreement was that the "upcoming member states"—Spain, Portugal, Greece—were given leeway to

[14]Beliczay– Szabó, *Az éghajlatvédelem gazdasági eszközei—Az emisszió-kereskedelem*. Levegő Munkacsoport, Budapest, 2003.

[15]On emissions trading in the EU, see e.g.: Grubb *et* Neuhoff, *Allocation and competitiveness in the EU emissions trading scheme: policy overview*. Climate Policy, Vol 6, No. 1., pp. 7-30; Klepper *et* Peterson, *The EU Emissions Trading Scheme. Allowance Prices, Trade Flows, Competitiveness Effects*. FEEM Working Paper No. 49.04.

[16]On how Burden Sharing works in the EU, see: S. Dessai et A. Michaelowa, "Burden sharing and cohesion countries in European Climate Policy: the Portuguese example." *Climate Policy* I (2001), pp. 327-341.

slightly increase their emissions, while richer member states would compensate for it. In parallel, in June 2000 the European Commission launched the first European Climate Change Programme (ECCP I).[17] The major goal of the ECCP I was to help EU member states achieve the Kyoto undertakings.[18] It is worth noting that the EU ETS was a result of the consultative process in the ECCP I working groups.

While the Kyoto Protocol was ratified by the EU in 2002, due to a lack of signatories it only entered into force in February 2005.[19] In 2005, the European Commission launched the second European Climate Change Programme (ECCP II).[20] The main task of ECCP II was to facilitate the genuine implementation of the priorities identified in the first program.

In March 2007, the Council authorized a proposal that the EU take a one-sided commitment to reduce emission by 20% by 2020. In March 2010, a new EU summit was organized to ratify the new energy action plan to come into force in 2010.

Table 1 shows the burden sharing within the EU.[21]

The European Commission evaluated the CO_2 emission of the member states to be 3457 Mt for 2010.[22]

---

[17]The ECCP is a multi stakeholder consultative program, bringing together various interests and expertise. It works in working groups forming forums from a range of policy sectors, industries and implementing hubs.

[18]The program was streamlined with the main EU strategy environmental program, the 6th Environmental Framework Programme (2002-2012).

[19]The European Union assumes an obligation according to the Protocol to reduce the emission of greenhouse gases of the year 1990 by average 8% by the time of 2008-2012. By applying the method of bubble mechanism and burden sharing the EU gained right to redistribute the committed 8% among the Member States because it signed the Protocol as an independent contractor.

[20]See the Commission's site: http://ec.europa.eu/clima/policies/eccp/second_en.htm.

[21]2002/358/EC: Council Decision of 25 April 2002 concerning the approval, on behalf of the European Community, of the Kyoto Protocol to the UNFCCC and the joint fulfilment of commitments hereunder. OJ L 130, 15.5.2002., 1-3.

[22]Communication from the Commission to the Council and the European Parliament—A Community strategy to promote combined heat and power (CHP) and to dismantle barriers to its development, COM(97) 514 final, http://www.resource-saver.com/file/toolmanager/O105UF456.pdf, accessed on August 23, 2011.

Table 1. The reduction of emission assumed in the Kyoto Protocol (% emissions of 1990 level in 15 member states of the European Union)

Belgium	92.5
Denmark	79
Germany	79
Greece	125
Spain	115
France	100
Ireland	113
Italy	93.5
Luxembourg	72
Holland	94
Austria	87
Portugal	127
Finland	100
Sweden	104
United Kingdom	87.5
European Union	92

The 2003/87/EC Directive

At the outset, and during the discussion in Kyoto, the EU refused to accept the emission quota trading system, but later radically modified its position. In 2001 the European Commission adopted a directive on the subject of the obligatory trading system of greenhouse gases. After the European Parliament's amendments in 2003, the EU legislation resulted in the above act.[23] Briefly, the directive defines the upper limit of emissions, sets out the theory of free assignment for most emissions, prescribes obligatory participation by the parties concerned, and assures legal enforcement of the derogations and fines for those who do not meet requirements.[24]

---

[23]For a more extensive assessment of the directive, see e.g.: J. Wettestad, "The Making of the 2003 EU Emissions Trading Directive: An Ultra-Quick Process due to Entrepreneurial Proficiency?" *Global Environmental Politics*, Volume 5, Number 1, February 2005, pp. 1-23.

[24]Directive 2003/87/EC of the European Parliament and of the Council of 13 October 2003 establishing a scheme for greenhouse gas emission allowance trading within the Community and amending Council Directive 96/61/EC. OJ L 275, 25.10.2003., pp. 32-46.

Table 2. Emission Permit vs. Emission Quota

Emission permit	Emission quota
Permit	*Allowance*
To the site, registered to the company	1t CO_2 emission
Not negotiable	Negotiable
The company cannot work without this	The licensee cannot emit CO_2 without it
It contains monitoring, return and certification fee	Standardized definition in the European Union
It obliges the licensee to present the same quantity of quotas in accordance with the annual CO_2 emission	All EU member states are obliged to accept
It does not contain emissions limit value	Any legal or natural person can buy it
	It is launched and can be cancelled by a national authority
	It has to be registered in a national register[26]

The directive at first was limited to CO_2 emissions, but later in 2004—and then again in 2006—the Commission proposed the extension of the directive to all greenhouse gases identified in the Kyoto Protocol. In addition, the directive obliged the member states to ensure greenhouse gas emissions permits for a certain group of companies, which permits each licensee the quantity of emission quotas in accordance with their greenhouse gas emissions. The emissions permit was a commitment and right for the given companies to trade and exercise accordingly. The permit also authorised the licensee to benefit from the emission quotas defined by the country. The emissions permit was not negotiable; it applied to the specified site and/or company. The emissions quota was a voucher with which the emitter could certify to the authorities.[25] Table 2 shows the difference between emission permit and emission quota.

The member states had to prepare an allocation plan before the period of trading ratified by the Commission. The plan had to contain the total quantity of quotas and the way in which the quota will be dis-

[25]Lesi– Pál, *A széndioxid emisszió kereskedelem elméleti alapjai és Európai Uniós szabályozása*, PM Kutatási Fejezetek 11. szám, 2005, http://www.rekk.eu/images/stories/letoltheto/pm-11.pdf?14c7e2ee2520855d5ac98ec049c29945=jwrvoavs, accessed on August 23, 2011.

[26]Ibid.

tributed. During the first regulation period the member state could auction 5% of the quotas to the companies. In the second (Kyoto) period this proportion was raised to 10%. New entrants were treated specially, through a reserve system. To track emissions allowances, each member state had to have its own national allowance registry. It was also possible to team up to combine registries. To harmonize monitoring, the Commission would operate a computerized independent transaction log that served as a centralized clearinghouse to verify allowance transfers between national registries.[27]

The total quantity of quotas was to be commensurable to the quantity assumed by the country in the Kyoto Protocol.

A portion of this national target had to be assigned to the installations participating in the EU ETS, with the remainder of the national target available for emissions outside the EU ETS. As Kruger and Pizer (2004) explained, this suggested a three-step process. "First, Member States must decide how much of their allowable Kyoto emissions will be assigned to the sectors included in the EU ETS [...]. Next, Member States may devolve this national ETS target into targets for each of the sectors included in the program. Finally, national program administrators must develop methodologies to allocate these sectoral targets to individual installations."[28]

If the licensee emitted more quotas than allowed, the penalty was €40 per tonnes exceeded during the first period, which rose to €100 per tonnes during the second. The missing quantity of quotas (that was not given to the authority) had to be compensated the following year. It is worth noting that monitoring was not enforced strongly as member states could decide about fraudulent reporting, late reporting, etc.

The ETS was the first trading system that allowed the trading with emission rights between countries. The countries divided the maximum emission value defined at national level into free tradable units, which were to be sold or rented to certain companies. Hypothetically, the quota market could emerge if the degree of emissions is higher than the

[27]See also: Kruger *et* Pizer, *The EU Emissions Trading Directive: Opportunities and Potential Pitfalls*. Discussion Paper 04-24, Resources for the Future, 2004, p. 5.

[28]Kruger *et* Pizer, *The EU Emissions Trading Directive: Opportunities and Potential Pitfalls*. Discussion Paper 04-24, Resources for the Future, 2004, p. 4.

one defined in advance. This way those countries that emit more greenhouse gases can buy these units from those that emit less. We should note that there is a possibility of emissions trading between companies within countries, but between countries and countries as well.[29]

Directive 2003/87/EC also set out the possibility of linking trading schemes with parties that have ratified the Kyoto Protocol. Countries that link with the EU ETS will have their allowances recognized in the EU system on the basis of a bilateral agreement between the European Union and that country. With this many issues were raised, of which up until now not many were resolved.

The EU ETS

The EU ETS (European Union Emissions Trading System) is an emissions trading scheme established by the European Union, which aims to offer incentives to reduce emissions of greenhouse gases economically.[30] Within this system heavy duty cal cars, oil refineries, coke ovens, metallurgic and steel plants, cement and lime-kilns, glass and building material factories and paper-mills can emit CO_2 only with permission.[31]

Initially the scheme involved some 11,000 industries that account for approximately 40% of greenhouse gas emission of the EU. Member states defined the quotas in the National Allocation Plans. One emissions unit was equal to 1 tonne of CO_2. Beside this, the so-called Effort Sharing (ES) system was established that controls the industry sectors, which produce lower level of emissions, e.g. transportation, building industry, agriculture and waste industry and furthermore the small emitters like households, small and medium sized companies and the ones from the supply sector. These "leftovers" would produce

[29]See also in: J. Lindmayer, *Emisszió kereskedelem a koppenhágai klímakonferencia tükrében*, 12 January 2010, http://www.biztonsagpolitika.hu/?id=16&aid=820&title=emisszio-kereskedelem-a-koppenhagai-klimakonferencia-tukreben, accessed on August 23, 2011.

[30]For an evaluation of the ETS, see, e.g.: Ellerman-Kuchner, „The European Union Emissions Trading Scheme: Origins, Allocation, and Early Results," *Review of environmental economics and policy*, Vol 1, No 1, 2007, pp. 66-88.

[31]See also: D. Fazekas, *Szén-dioxid piac az Európai Unió új tagállamaiban*, PhD study, Budapest, 2009, http://phd.lib.uni-corvinus.hu/415/3/fazekas_dora_thu.pdf, accessed on August 23, 2011.

some 55% of emission of the European Union.[32] As one of the most significant changes, part of these sectors will become a component of the ETS in the third trading period.

In the first two trading periods (2005-2012) the main part of the emission units were allocated free of charge to the companies. At the beginning of the third trading period, on 1 January 2013 the ETS system changes significantly. The rules become more aligned; and the way the system works becomes more computable.[33]

Table 3 shows the ETS emission units of each member state in the period of 2005-2012.

As already mentioned, EU member states assumed a commitment to reduce the emission of greenhouse gases by 8%. Later this common aim was transformed into national goals in the 2002/358/EC Council Decision. Only the national commitments apply to the 12 states that joined the European Union between 2004 and 2007 (except Cyprus and Malta, to whom these kinds of commitments were not applied).

The ETS now operates in the 27 EU member states plus Iceland, Liechtenstein and Norway. It covers a large quantity of CO_2 emissions from various installations such as power stations, combustion plants, oil refineries and iron and steel works. Newcomers are factories making cement, glass, lime, bricks, ceramics, pulp, paper and board. Nitrous oxide emissions from certain processes are now covered as well.[34]

---

[32]See also in: J. Lindmayer, *Emisszió kereskedelem a koppenhágai klímakonferencia tükrében*, 12 January 2010, http://www.biztonsagpolitika.hu/?id=16&aid=820&title=emisszio-kereskedelem-a-koppenhagai-klimakonferencia-tukreben (23 August 2011).

[33]The EU against the climate changes—The emission trading system of the EU, European Communities, 2009, http://ec.europa.eu/clima/publications/docs/ets_hu.pdf, accessed on August 23, 2011.

[34]There is a vast number of analyses about the ETS. To understand the economics behind it, see, e.g.: Neuhoff *et al.*, *Allocation, incentives and distortions: the impact of EU ETS emissions allowance allocations to the electricity sector*, 19 May 2006. Available at: http://www.dspace.cam.ac.uk/bitstream/1810/183627/1/eprg0618.pdf, accessed on September 30, 2011.

Table 3. ETS Emission Units of Each Member State, 2005–2012.

		2005-2007		2008-2012	
Country	Kyoto target value (% change)	Allocated CO_2 emission unit (mton/year)	Share from the ETS (%)	Allocated CO_2 emission unit (mton/year)	Share from the ETS (%)
Austria*	-13	33.0	1.4	32.3	1.5
Belgium*	-7.5	62.1	2.7	58.0	2.8
Bulgaria	-8	42.3	1.8	42.3	2.0
Cyprus	-	5.7	0.2	5.2	0.3
Czech Republic	-8	97.6	4.2	86.7	4.2
Denmark*	+21	33.5	1.4	24.5	1.2
Estonia	-8	19	0.8	11.8	0.6
Finland*	0	45.5	2.0	37.6	1.8
French*	0	156.5	6.8	132.0	6.3
Germany*	-21	499	21.7	451.5	21.6
Greece*	+25	74.4	3.2	68.3	3.3
Hungary	-6	31.3	1.4	19.5	0.9
Ireland*	+13	22.3	1.0	22.3	1.1
Italy*	-6.5	223.1	9.7	201.6	9.7
Latvia	-6	4.6	0.2	3.4	0.2
Lithuania	-8	12.3	0.5	8.6	0.4
Luxembourg*	-28	3.4	0.1	2.5	0.1
Malta	-	2.9	0.1	2.1	0.1
Netherlands*	-6	95.3	4.1	86.3	4.1
Poland	-6	239.1	10.4	205.7	9.9
Portugal*	+27	38.9	1.7	34.8	1.7
Romania	-8	74.5	3.2	73.2	3.5
Slovakia	-8	30.5	1.3	32.5	1.6
Slovenia	-8	8.8	0.4	8.3	0.4
Spain*	+15	174.4	7.6	152.2	7.3
Sweden*	+4	22.9	1.0	22.4	1.1
United Kingdom*	-12	245.3	10.7	246.6	11.8
Liechtenstein	-8			0.2	0.0
Norway	+1			15.6	0.7
Total		2298.5	100 %	2086.5	100 %

* indicates the states that actually made commitments.

Effects of the Third Trading Period (ETS after 2012)

The Kyoto Protocol was affirmed by 176 countries plus the EU, and finally came into effect in February 2005.[35] Different values are applied to the parties in the different annexes but the general aim for the period of 2008-2012 is to reduce CO_2 emission by 5% in relation to the greenhouse gas emission level of 1990.

Because at the outset trading covered only approximately 40% of greenhouse gases, EU CO_2 trading was supposed to increase and become more heterogeneous over time. Since 2005 the ETS has enlarged—with the entry of further countries and the emergence of the so-called gas specific option (joining the ETS of coal-burning power plants of efficiency lower than N_2O and 20MW)[36]—and the Commission and the Member States have decided that for the start of the third trading period in 2013 there is a need for clearer rules and better functioning trading market for the effective reduction of emissions.

The Third Period of Emissions Trading in the EU (2013-2021)/ The New Emissions Trading Constitution: Directive 2009/29/EC

According to the directive that reformed EU emissions trading in 2009, emissions units launched after January 1, 2013 are valid for eight years. From 2013, the quantity of annual emissions units in the EU will be reduced in a linear fashion in accordance with the safety decisions taken concerning the ratified national allocation plans between 2008 and 2012, and as compared to the average annual emissions units of the member states—in line with a 1.74% linear value.[37]

According to the new rules, in the third period there will be no strongly heterogeneous national distribution plans: the Commission will control an integrated European quota distribution system. This means that the EU enforcer will help to set up and closely monitor how the distributions of allowances proceed in the member states.

---

[35] The United States has not signed the Protocol.

[36] See also MIT Interim Report, *The European Carbon Market in Action: Lessons from the First Trading Period*, Report No. 162, June 2008, p. 22.

[37] Article 9, Directive 2009/29/EC of the European Parliament and of the Council of 23 April 2009 amending Directive 2003/87/EC so as to improve and extend the greenhouse gas emission allowance trading scheme of the Community, OJ L 140, 5.6.2009, pp. 63-87.

Another new element of the third trading period is that the ETS will encompass new industries: among others,[38] glass and porcelain manufacturers and the aluminium and ammonium manufacturers become part of the system.

In keeping with the new national reporting system, the Commission can use the available information to propose the monitoring of the whole trading system to the Council and the Parliament, to better clarify and understand the carbon market.

The overall aim of the allocation of quotas—with the modification of numerous details—concerning the trading period starting from January 1, 2013, is the effective evaluation of environmental protection activities, appreciation of the investments of companies in more efficient emission, and the reduction of environmental pollution by retaining European competitiveness.

By 2020 the EU intends to reduce the emission of greenhouse gases by 20% compared to the level of 1990. As part of this aim, the Commission is constantly monitoring the developments in this field and as one of the last actions; it launched a public consultation on further reducing industrial gas emissions of fluorinated gases in September 2011.[39]

As the Directive only sets targets at the EU level, the Member States will have to implement the regulations of the directive in national legislation by December 13, 2012.

Air Transport: A Newcomer to the European Trading Scheme

Taking into consideration all modes of transportation, airplanes emit the most carbon dioxide. According to the Stern Review, this sub-sector produces 12% of the total emissions in the transportation sector,[40] significantly after road transport (78%).[41] The Commission

[38]The detailed list of the new activities can be found in directive 2009/29/EC, appendix I.

[39]See for further details: IP/11/1078, at: http://europa.eu/rapid/pressReleases Action.do?reference=IP/11/1078.

[40]Annex 7.c, *Stern Review on the Economics of Climate Change*, 2006, See: http://webarchive.nationalarchives.gov.uk/+/http://www.hm-treasury.gov.uk/d/Transport_annex.pdf.

[41]According to the review, the fact that air transportation was the most developing

estimates that an aircraft doing a return flight between Brussels and New York generates about 800 kg of CO_2 per passenger. Furthermore, air traffic represents around 10% of greenhouse gas emissions covered by the EU ETS.[42]

In 2005, when the first ETS trading period started, none of the 11,000 companies was dealing with air transportation. The second trading period—according to the original plan of the Commission—contained the quotas allocated for treatment of emission coming from air transportation: from 2011 for the airlines within the EU, from 2012 for the international lines as well. According to the proposal adopted in January 2009,[43] from 2012 companies participating in European air transportation have to reduce emission to 97% of the level of 2005, then from 2013 to 95% of the latter.[44] The commitment applies to all national and international aviation emission of CO_2: airplanes entering European airspace have to buy ETS credits for the whole distance, not only for passage in Europe.

According to the rules, 15% of the aviation emissions units have to be allocated via auction. As is stated in point 3 of the directive—although the detailed rules are made by Member States—the resulted income should be used "[...] to tackle climate change in the EU and third countries, inter alia, to reduce greenhouse gas emissions, to adapt to the impacts of climate change in the EU and third countries, especially developing countries, to fund research and development for mitigation and adaptation, including in particular in the fields of aeronautics and air transport, to reduce emissions through low-emission transport and to cover the cost of administering the Community scheme."[45]

---

industry between 1990 and 2002 means that the industry could achieve some 9GtCO_2 emissions by 2030.

[42]See also in: IP/11/1077 of 26.9.2011.

[43]Directive 2008/101/EC of the European Parliament and of the Council of 19 November 2008, amending Directive 2003/87/EC so as to include aviation activities in the scheme for greenhouse gas emission allowance trading with the Community, OJ L 8, 13.1.2009.

[44]According to the Directive: "*[...] for each subsequent period, the total quantity of allowances to be allocated to aircraft operators shall be equivalent to 95 % of the historical aviation emissions multiplied by the number of years in the period.*" Directive 2008/101/EC, Article 1, p. 8.

[45]Directive 2008/101/ EC, Article 1, point 4, p. 8.

The carbon emissions of the aviation sector became the subject of monitoring activity in 2010. Following the exercise, more than 900 airlines applied for free allocation of emissions units. Consequently, at the end of September 2011 the Commission published the benchmark values for airlines. These values will serve as the basis for the allocation of greenhouse gas emission allowances free of charge to about 900 aircraft operators. There are two benchmarks: one has been calculated for the trading period in 2012, while the other for the ETS III. In 2012, airlines will receive about 0.6797 allowances, and between 2013 and 2020 an airline will receive somewhat less.[46]

The international response to these measures has not been friendly, mainly because overseas it is considered an overreach into the affairs of operators not based in Europe. Operators in the U.S., China and Russia argue that the interference in the aviation sector through the ETS violates the Chicago Agreement (1944). A case is currently being heard before the European Court of Justice.[47]

The Entrance of the Aluminum Industry to the ETS—An Energy-Intensive Example of the New Regulation

Reports of the effects of ETS on the aluminum sector in previous years emphasized how unfair it was for this sector still to be part of ETS after 2012.[48] The studies argued that eventually certain energy-sensitive industries—like aluminum—will have to leave Europe and search for another territory where regulations of climate policy and CO_2 trading are more advantageous, so as to be able to keep up with their competitors globally. Yet, statements like this are somewhat misleading because an IAE report declared that the European aluminum

[46]See: Commission Decision of 26 September 2011 on benchmarks to allocate greenhouse gas emission allowances free of charge to aircraft operators pursuant to Article 3e of Directive 2003/87/EC of the European Parliament and of the Council, OJ L 252, 28.9.2011, pp. 20-22.

[47]See: C-366/10. See at: http://eur-lex.europa.eu/LexUriServ/LexUriServ.do?uri=OJ:C:2010:260:0009:0010:EN:PDF (10 September 2011).

[48]See e.g. Reinaud, J. (2008). *Climate policy and Carbon Leakage: Impacts of the European Emissions Trading Scheme on Aluminium.* IEA information paper, October 2008. Available: http://www.iea.org/papers/2008/Aluminium_EU_ETS.pdf.

industry has long been at a disadvantage in comparison with similar production of other countries since 1999.[49]

Despite constant European-led efforts by the aluminum lobby, beginning in 2013 primary and secondary aluminum producers will become part of the CO_2 trading system. According to directive 2009/29/EC, companies dealing with primary and secondary aluminum production will enter the trading system if they apply stokers with higher input power above 20 MW.

Aluminum producers are, however, not the only traditionally energy-sensitive new entrants: precious metal production, ceramics, the glass industry and petrochemicals are among the new traders relating to directive 2009/29/EC.

The entry into the scheme of these new sectors of industry could potentially result in carbon—or competitiveness—leakage and greater calls for the expatriation of such energy-sensitive industries from the European region. Will the relocation of these industries take place if trade patterns and uneven carbon constraints continue on the present course?

At the time of the final stages of this study there is no evidence available that supports the argument of the aluminium industry. The cited IAE report notes that "[...] about 85% of Europe's primary aluminium imports originate from eight countries: Norway, Russia, Mozambique, Brazil, Iceland, United Arab Emirates, Canada and South Africa. At present, it costs more to produce a tonne of primary aluminium in Europe than in many other regions. However, this was already the situation in 1999, prior to the introduction of a carbon cost in the EU. The carbon constraint is obviously only one element in this picture, as higher electricity prices prevailed before the introduction of the ETS [...]."[50]

[49]Reinaud, J. (2008). *Climate policy and Carbon Leakage: Impacts of the European Emissions Trading Scheme on Aluminium.* IEA information paper, October 2008, p. 3.

[50]Ibid.

Changing the Rules on Auctioning

In the first two trading periods, auction affected only a limited quantity of quotas, while in the third period it becomes the main rule of allocation.[51] According to the plans, the Commission aims to put up for sale half of the total quantity of emissions units.[52] The executive authority took action to leave behind the secondary coordinating rule considering earlier allocations. It now intends to define the rules ex ante, with stricter control over the allocation of the specific member states ex post.

The issue of individual allocations may raise the possibility of illegal national aid. Article 107 of the Lisbon Treaty and the related secondary rules prohibit illegal state aid that can put a company into an advantageous situation and distorts competition, while could influence trading among member states. A certain quantity of CO_2 allocated in an inadequate way may serisly influence the expenses of a company, such as electricity production or petrochemicals, because it may bring about an extremely advantageous situation for other companies. It is even more dangerous if companies are entitled to receive large quantities of free quotas without a just environmental effect analysis and then have the possibility for trading them freely.

The Rules of the Auction Processes for the Third Period are Regulated by Regulation 1031/2010/EU[53]

The auction process is designed on the basis of open, clear and integrated principles. The primary platform will be the European platform, but it does not mean that member states cannot operate their auction platforms. In July 2011 member states committed themselves to the use of a particular auction mechanism, and confirmed in a

[51]About the auction rules of the second period see e.g. Hepburn *et al.*, 2006. *Auctioning of EU ETS phase II allowances: How and Why?* Climate Policy 6: 137-60.

[52]During the first probation period the parties taking part in allocation were the Commission, the governments of Member States and larger companies. The allocation and the individual distribution were executed via the National Allocation Plans (NAP), while micro decisions i.e. the portion of the quotas for the companies were set by the industrial lobby.

[53]Commission regulation No 1031/2010 of 12 November 2010 on the timing, administration and other aspects of auctioning of greenhouse gas emission allowances pur-

common agreement the auctioning of 120 million allocation units in 2012.

The auction system is based on the trading of the five day-long valid spot products.[54]

The Commission states that "[...] auctions shall be carried out through an auction format whereby bidders shall submit their bids during one given bidding window without seeing bids submitted by other bidders."[55] The auction has to be terminated in one day, and a two-hour-long period is necessary between each bid, while one period may take at least two hours.

Calculation of the quantity auctioned is a result of a rather complicated method. It is conceived in Article 10 (2) in the following way: "The volume of allowances covered by Chapter III of Directive 2003/87/EC to be auctioned in 2013 and 2014 shall be the quantity of allowances determined pursuant to Articles 9 and 9a of that Directive for the calendar year concerned, less the allocation free of charge provided for in Articles 10a (7) and 11(2) of that Directive, less half of the total volume of any allowances auctioned in 2011 and 2012." It goes on, "[...] the volume of allowances to be auctioned each calendar year as from 2013 shall be based on the Commission's determination and publication pursuant to Article 10 (1) of that Directive of the estimated amount of units to be auctioned or on the most recent amendment of the Commission's original estimate as published by January 31 of the preceding year."[56] Article 18 defines the circle of bidders.

Generally, it can be said that the regulation safely sets up auction trading. It controls the auction platform, defines duties of the auction controller and the participants, and furthermore regulates the integrate platform of the two- and five-day-termed futures. It also defines the platform of forwards and futures. The Commission expects the first effects of the regulation by December 31, 2014.

---

suant to Directive 2003/87/EC of the European Parliament and of the Council establishing a scheme for greenhouse gas emission allowances trading within the Community, OJ L 302, 18.11.2010, pp. 1-41.

[54]The regulation allows forwards and futures as well with specified conditions.

[55]Article 5 of the Directive 1031/2010/EU.

[56]Article 10 (3) of the Directive 1031/2010/EU.

The Price-Correctional Safety Clause

One of the most important aims of the allocation policy and homogenization of auctions is to avoid carbon quota price volatility. Therefore the modified directive introduces a so-called crisis commission coordinated by the European Commission.[57]

This commission type will be set up if the prices of emissions units are higher than triple the average European carbon market price of the previous two years for six consecutive months.

The directive declares that "[...] if the price evolution referred to in paragraph 1 does not correspond to changing market fundamentals, one of the following measures may be adopted, taking into account the degree of price evolution:

(a) a measure, which allows Member States to bring forward the auctioning of a part of the quantity to be auctioned;

(b) a measure, which allows Member States to auction up to 25% of the remaining units in the new entrants reserve."[58]

The number of allowances set aside for new entrants by the 25 member states for the first trading period of the EU ETS was about 3 percent of the total. The percentage set aside varied, from as little as 0.4% in Poland to 26% in Malta. The theory of distribution was on a "first-come-first-served" basis.[59]

Whether price volatility of carbon markets—highly fluctuating prices and no banking allowed in the first and second trading period—will be targeted efficiently by this safety clause remains to be seen. It is, however, likely that a more comprehensive set of data about the allowances through monitoring reports, a standardized national legal framework and avoiding of free over allocation of credits will calm

[57]Directive 2009/29/EC of the European Parliament and of the Council of 23 April 2009 amending Directive 2003/87/EC so as to improve and extend the greenhouse gas emission allowance trading scheme of the Community, OJ L 140, 5.6.2009., pp. 63-87, Article 1, point 29

[58]Ibid.

[59]Ellerman-Kuchner, *The European Union Emissions Trading Scheme: Origins, Allocation, and Early Results*. Review of environmental economics and policy, Vol 1, No 1, 2007, pp. 75.

down frustration between companies and sectors. Consequently, trading markets should be better suited for doing business and windfall profits are not expected to aggravate participants.

The NER300 Programme

The so-called NER300 programme[60] is not directly part of the rules regulating the European emissions trading but the first apparent effects of it will probably coincide with the start of the third trading period. The program, which started officially in November 2010, was named after part of the units—300 million CO_2 emissions units—for new ETS entrants. Its aim is to establish projects operating with low CO_2 emission technologies from the monetizing of the units, which approach and motivate effective and low CO_2 emission.[61]

The projects advance renewable energies and innovative carbon-dioxide storage. Complementary supporting projects had to be brought in by member states to the European Investment Bank by May 9, 2011 after having been declared adequate and worth being supported.[62]

According to the initial plans from 2010, the Commission will co-finance approximately 8 projects for carbon-dioxide storage and 34 projects for supporting renewable energy.[63] Commission co-financing means that projects that have to be co-financed can be financed—of the relevant costs—with 50% coming from other complementary resources.[64] Co-financing may be found from the Cohesion and Structural funds or in the European Energy Programme for Recovery

---

[60]NER: New Entrance Reserve 300.

[61]The programme was established by the Commission Decision C(2010) 7499 final, available at: http://ec.europa.eu/clima/funding/ner300/docs/c_2010_7499_en.pdf, accessed on September 26, 2011.

[62]The Bank monitors the projects and is responsible for selling 300 million emission allowances.

[63]Financial distribution of each field of renewable energy see: http://europa.eu/rapid/pressReleasesAction.do?reference=MEMO/10/549&format= HTML&aged=0&language.

[64]Co-financing may be combined with *loan financing* provided under the Risk-Sharing Finance Facility (RSFF) set up by the Union and the European Investment Bank (EIB). See also: Commission Decision C(2010) 7499 final, (5).

(EEPR). The projected costs are strictly accounted for, and only costs with extra investment can be included. Thus fixed charges, preference taxes and reducible costs by outgoing from ETS cannot be accounted for. The final financing value of the projects—because of the 300 million emission units for sale—depends on the actual market price of the emissions monetized. The Commission plans to decide on the first projects by the second half of 2012.

Projects that may receive financing must "[...] make use of technologies which are innovative in relation to the state of the art in the key sub-streams for each technology. Those technologies should not yet be commercially available, but sufficiently mature to be ready for demonstration at pre-commercial scale. They should have reasonable prospects of successful demonstration, taking into account that technological risks are inevitable, and the proposed scale of demonstration should be such that no significant additional problems are to be expected from further scaling up."[65]

At the same time, projects must be ones that may be recreated on other occasions and other places. One-site innovations will not be funded as they do not have a broad European effect.

The Third Trading Period and Energy Production

The preparation of the EU emissions trading system for 2013 and beyond is still in progress. During this process member states can reveal their national interests along certain conditions according to the modified directive 2003/87/EC. This means that a member state can, with its auction rights, support a special industry permanently by allocation of free emission units. This special industry is the electricity industry.[66]

Directive 2003/87/EC—modified by directive 2009/29/EC—lays down that in energy production it is not possible to allocate free quo-

[65]Commission Decision C(2010) 7499 final, (10).

[66]This industry is special because in the first two trading periods almost all the shortages of emission quota allowances were allocated to it. If a country had a lack of quotas, these were generally allocated to energy production. The reason for this is that it is the market where there is the least competition from outside Europe. See also: MIT Interim Report, *The European Carbon Market in Action: Lessons from the first trading period*, Report No. 162, June 2008, pp. 11.

tas unless energy is produced from recycled gas. Besides, this "[...] Member States may give a transitional free allocation to installations for electricity production in operation by 31 December 2008 or to installations for electricity production for which the investment process was physically initiated by the same date, provided that one of the following conditions is met:

(a) in 2007, the national electricity network was not directly or indirectly connected to the network interconnected system operated by the Union for the Coordination of Transmission of Electricity (UCTE);

(b) in 2007, the national electricity network was only directly or indirectly connected to the network operated by UCTE through a single line with a capacity of less than 400 MW; or

(c) in 2006, more than 30% of electricity was produced from a single fossil fuel, and the GDP per capita at market price did not exceed 50% of the average GDP per capita at market price of the Community."[67]

Free allocation is possible only if the Commission accepts the proposed project of the member state, which also contains the diversification of actual source of energy. Emissions units allocated by the ratified emission plan have to be deducted from the member state's quotas in auction.

In the third trading period member states can exempt the so-called small emission institutions from the trading system with certain conditions, restricted to cutting administrative costs. These institutions are the ones of which rated power input in the last three years did not exceed 35 MW and the then actual power output was lower than 25,000 tonnes CO_2. These institutions then cannot participate in free allocation because they do not even have restitution liability. If the output exceeds the actual trading value, the institution returns to ETS and cannot leave even if in the meantime it would become entitled to leave.

The chosen small institutions have to meet several requirements; first of all, the member state has to prove to the Commission that it

[67]Directive 2009/29/EC.

Table 4. Differences in the Emissions Trading System of the EU in Periods I and II and III

Period I and II	Period III
National total quotas	EU quota
Fix quotas	Fix quotas that reduces annually
3 and 5-year trading period	8-year trading period
Determined auction (<4%)	Significant auction – constitution
Free allocation for industry + energy producers	Temporary free allocation for emissions concerning industry and heat (not for energy production)
Free allocation based on institutional emission	Free product allocation based on specified emissions
Free allocation based on past emissions	Free allocation based on referential values
Argument:	Argument:
Directive 2003/87/EK	Amended directive 2003/87/EK
National allocation plans	Community Implementing Measures (CIM)
EC decisions about the NAP plans	National Implementing Measures (NIM)
National allocation decisions	

introduces measures equal to the ETS for reducing environmental pollution. The Commission has three plus six months to judge the declared demands of the member states. Emissions allowances and reports have to carry on as they are. Furthermore, the fees for control and certain administrative services do not change. Yet, a new cost emerges related to getting used to the new system. (see table 4)

An important novelty of the ETS III system is the linking of the system with other emissions trading systems.

EU-U.S. Relations with Respect to Carbon Trading

There is a deep-seated difference between how the EU and the U.S. approach climate change issues, including emissions trading. Does the much advocated consensus between the parties that climate challenge can only be tackled jointly exist at all? Or are we talking about two completely different protectionist approaches to national carbon markets?

The Bush administration set the United States back with respect to where the EU stood on climate change and sustainable energy.[68] President Obama tried to change this to some extent. First, a cap and trade bill nearly came to a Senate vote in 2008, amid a flurry of proposals clearly marking territory for a real, rather than symbolic, approach to the issue. At the outset, President Obama made it clear that he intended to change America's approach to climate policy both at home and abroad.[69]

> The cap and trade system is a scheme of emissions trading whereby the central administration defines the total amount of the tradable quantity—cap—that the designated sources can emit over a certain period of time, while the designated market players may trade them in order to cover their needs or to win extra profit by functioning eco-friendly.[70]

It seems that the EU implemented significant preliminary U.S. approaches, which were then abandoned by previous U.S. administrations. Most of all, Brussels established the world's first international carbon trading scheme and without delay, learned from the mistakes of the first-mover. However, the EU has chosen a significantly different way of incentivizing the participants. Whereas in the U.S. on-time allocation would take place—in the context of the U.S. SO_2 and NO_x programmes—the EU ETS adopted a so-called sequential approach. Allocation plans were accepted for one commitment period, with recurring negotiations about the allocation for the subsequent period. As Neuhoff et al. put it: "Although consistent with the iterative nature of international emission reduction negotiations, this allocation approach can have significant implications to efficiency of the market compared with one-off allocation. [...] it creates perverse incentives for CO_2 intensive plants to remain in operation in order to receive free-

[68]At last, it signed the Bali Roadmap under the UNFCCC.

[69]See also in: J. Anderson, *The Carbon And Credit Crisis: Challenges and Opportunities for Transatlantic Relations*, (IEEP), April, 2009. Available at: http://www.boell.eu/downloads/carbon_and_credit_crisis_paper.pdf, accessed on September 13, 2011.

[70]Particularities of the system: opt-ins, set-asides, offsets. For a longer explanation, see, e.g.: Aulisi *et al*: *Greenhouse gas emissions trading in U.S. States. Observations and lessons from the OTC NOx Budget Program.*, WRI, 2005. p. 5.

allocations, even if closure or replacement is socially more efficient. In addition, firms might invest in and operate more carbon intensive technologies if they anticipate that future allocations of allowances will be proportional."[71]

Probably not all the advantages of such a system have been identified since the entry into force of the ETS in 2005. In addition, Europe is without a transatlantic partner in emissions trading. There may be many reasons for this, of which global views on climate change and the participatory nature of greenhouse gas reductions are prominent. The Copenhagen Accord from 2009—in which the U.S. made its first commitments—can only be considered a start, as it has no legally binding effect.[72]

The U.S. started to reduce air pollution by regulatory actions—e.g. the Clean Air Act—as early as in the 1970s.[73] A later amendment of this first major climate change related legislation in 1990 introduced the possibility of emissions trading.[74] The long awaited—and spectacularly soon-disappearing—Climate Bill was watered down in 2010. What were the reasons behind the immediate halt?

One of the significant internal issues in the U.S. is how to limit the cost of compliance while trying to maintain long-term environmental and eco-environmental stability. The other is—as has been pointed out by several authors—how to engage developing countries so as to persuade them to seriously limit their emissions.[75] This latter issue is of particular interest in relation to China and India, two of the main trading partners of the U.S., and are on the top of the developing

[71]Neuhoff *et al.*, *Allocation, incentives and distortions: the impact of EU ETS emissions allowance allocations to the electricity sector*, 19 May 2006, pp. 3-4. Available at: http://www.dspace.cam.ac.uk/bitstream/1810/183627/1/eprg0618.pdf, accessed on September 30, 2011.

[72]The US agreed to set emissions at the level of 83% of its emissions in 2005. See: http://unfccc.int/resource/docs/2009/cop15/eng/l07.pdf, accessed on September 15, 2011.

[73]P.L. 88-206., Effective of December 17, 1963: Significant amendments in 1970, 1977, 1990.

[74]104 Stat. 2468, P.L. 101-549.

[75]Whether the "safety valves" mechanisms would be a better choice to compensate compliance costs—yet to be seen. See, e.g. M.W. Wara and D.G. Victor, *A realistic Policy on International Carbon Offsets*, PESD WP No. 74., April 2008.

countries' list. Preferable reduction scenarios for these countries are necessary in order to drive competitiveness.

However, the political quarrel over how to reduce emissions seems to be purely theoretical in nature as the U.S. Congress has not ratified the Kyoto Protocol—and its flexible compliance mechanisms— and therefore has not undertaken any internationally binding measures in the direction of avoiding the negative effects of climate change. The exploration of the cap and trade systems in California and in the northern U.S. states (RGGI) points in the direction of an unwillingness on the part of federal lawmakers, as opposed to local willingness to create carbon trading systems.

Prof. Stern notes: "[...] The possible outcomes that need to be considered, which include major irreversible changes to the climate, are likely to be considerably beyond human experience hitherto. Such uncertainty over the scientific, economic and social consequences of climate change makes it especially challenging for international collective action to agree on greenhouse gas emission targets."[76]

At the dawn of the third trading period, important questions reemerge: How does the U.S. approach the third trading period in the EU, where also clearly U.S.-based air companies are involved in the must-reduce, must-comply, must-buy/sell trading scheme, as air traffic joins the scheme? Will the U.S. only aim to fight the EU legislation at international aviation forums? Does the U.S. plan to counteract with a federal scheme that would eventually have effect on European players too? Or will it move towards the question in a cooperative way and signs the Kyoto Protocol, coordinate with the EU and the developing world, while reducing its emissions by the required percentage?

Conclusions and Recommendations

The success of the third trading period in the EU depends on available information and the administrative costs related to trading of emissions, which the companies will have to face. Instability and volatility of the quota prices may be avoided if the Commission exam-

---

[76]N. Stern, What is the Economics of Climate Change?, *World Economics*, Vol 7, No 3, April-June 2006, p. 9.

ines on time the national requests and reduces the likely heterogeneous national allocations.

NER 300 may play an important role because a CO_2 or energy supporting project, which can be financed totally, could influence functional costs and trading positioning not only in the ETS but also in the regional trading systems.

As to the period after the Kyoto Protocol expires, the global situation regarding the regulation of greenhouse gas emissions is hardly predictable. The failure of recent negotiations (Copenhagen, Cancún) made it clear that, although states—more or less—share each other's opinion about importance of reducing global warming of the earth (precisely slowing the process), there are significant differences in their economic, political and social interests in doing something about it. Any efficient response to climate change "[...] must be based on an international understanding that its origins, impact, scale and urgency require action that is global and collective."[77]

The success of the third ETS trading period—the world's largest carbon market—largely depends on the third parties and external players: i.e. on non-EU countries and global stakeholders, too. Yet, the first internal success-test will be the accessibility to information that is required for trading for the parties, price volatility of the quotas and the opportunism of the industry in planning ahead. In parallel, the European Commission will have to speed up in analysing the national requests so that from an administrative point of view, there are no burdens to trade and the market is not sluggish. Market sensitive information and regulator's plans are to be accurately announced and made accessible. The monitoring of the non-legal requirement of the Member States concerning the reinvesting of the income from credit auctioning is a sensitive exercise, which must be reinforced. Given the uneasy relationship fuelled by the 2012/2013 ETS trading rules, this is a vital point also for EU-U.S. climate change related negotiations.

In turn, the U.S. will also have to rethink the constructive nature of the *Anti-ETS Bill*, which is likely to cause more quandary than benefit for the aviation industry on both sides of the Atlantic. A questionable patriotic response to an allegedly extraterritorial act of the EU seems

[77]Ibid.

to be rushed and economically counterproductive. It is advisable that the EU and the U.S. continue to search for the common tone in this regard, but foremost, the U.S. reassesses the Act.

In addition, because of the results of global greenhouse gas emissions is not predictable, especially after the expiration of the Kyoto Protocol, it is vital that solutions are finalized in the shortest amount of time and that they do not remain regional, but reach the global scale. Emissions reduction techniques aren't particularly valuable if they only come from one part of the world, whereas other regions benefit from the efforts of the more dedicated states. There is urgency but action must be unified, applicable and monitoring-proof. The global players must aim for a speedy compromise.

Chapter Four

Innovation in the Energy Economy: An Imperative for Transatlantic Cooperation

Mihaela Carstei

The world has an insatiable appetite for energy. The IEA forecasts energy demand to increase 36% between 2008 and 2035, which will require the utilization of all economically available energy resources for many decades.[1] $38 trillion of investment will be required simply to meet the projected energy demand.[2] The global economic crisis and accompanying economic downturn have made meeting this challenge exceedingly difficult.

In a carbon-constrained world, new technologies will be needed to facilitate a transformation of the energy industry to meet higher environmental standards. Restructuring current energy systems toward a far greater reliance on technologies with low or no carbon dioxide emissions is an immense challenge. Fossil fuels such as oil, coal, and natural gas together satisfy 81% of global energy demand and generate 69% of global anthropogenic greenhouse gas emissions.[3] Moreover, fossil fuels are expected to remain dominant in the global energy mix with a share of 74% of worldwide demand for energy in 2035.[4]

Energy poverty remains prevalent throughout much of the developing world. If we are to live in a 21st century more prone to peace than violence, the developed countries must move expeditiously to address the developing countries' needs for energy.[5] The availability,

[1] International Energy Agency, *Key World Energy Statistics: 2010.*

[2] International Energy Agency, "Cumulative investment in energy infrastructure, 2011-2035," available at http://www.iea.org/weo/Files/2011_EBC_Ministerial_Press.pdf

[3] International Energy Agency, *Key World Energy Statistics: 2010.*

[4] Ibid.

[5] CNA Corporation, "National Security and the Threat of Climate Change," 2007,

accessibility and affordability of energy is vital to the economic development that is required to alleviate global poverty, to reduce global tensions, and to address global environmental degradation. Without a radical change in policies in the developing and the industrialized countries, there will be about the same number of people without access to electricity (1.3 billion) or relying on non-commercial biomass fuels (2.7 billion) in 2030 as today. Furthermore, it will be so despite a relatively rapid growth in energy consumption in the developing world.[6]

The challenges of energy security, climate change and energy poverty are immediate and vast. As such, to ensure a sustainable energy future a dramatic transformation of the world's energy supplies and consumption patterns is required. This transformation will affect virtually all economic sectors, as the energy systems that power them still rely heavily on carbon-intensive fossil fuels. The development and deployment of a portfolio of low-carbon technologies is an essential component of the needed energy industry modernization and restructuring. This will necessarily involve greater utilization of non-carbon based energy as well as reduction of carbon dioxide (CO_2) emissions from fossil fuels. New technologies are also essential to managing the forecasted costs and ensuring reliability.[7]

While much of the world's expertise on the clean-energy technologies needed to address these challenges currently lies within the U.S. and the EU, neither Europe nor the U.S. will be capable of meeting these challenges in isolation. The transatlantic community has an opportunity to work together to foster innovation to revitalize languishing industries, accelerate the development of advanced technologies, and become an example for countries struggling to develop and implement appropriate policies that support and accelerate innova-

---

available at http://www.cna.org/sites/default/files/news/FlipBooks/Climate%20Change%20web/flipviewerxpress.html.

[6] International Energy Agency, "Energy for All: Financing Access for the Poor," Special early excerpt of the *World Energy Outlook, October 2011*, Available at http://www.iea.org/papers/2011/weo2011_energy_for_all.pdf.

[7] Richard Lawson, John Lyman, Mihaela Carstei, "A Shared Vision for Energy and Climate Change: Establishing a Common Transatlantic Agenda," The Atlantic Council of the United States, 2010.

tion. The world is looking to the developed countries to lead, and leadership by the transatlantic community is crucial.

Current Conditions Impacting Energy Innovation

On both sides of the Atlantic today there common agreement on the need to create a "global revolution"[8] in energy production and use. Various groups and individuals, including leading business and national security figures, political leaders and international organizations have all called for significant changes in energy production and consumption to meet the goals of sufficient, reasonably priced, and sustainable energy. Efficient and effective technologies, policies and regulations will be required to sustain economic growth throughout the world. This will not occur unless the transatlantic community moves in concert to increase the efficiency of energy use, and to develop and deploy the technologies required to meet the needs of both the developed and developing countries.[9]

Government policies and energy prices have an important impact on the pace of development and deployment of new technologies.[10] While all energy technologies must compete in the marketplace in terms of cost, reliability and ability to attract capital, the recent financial crisis and economic downturn, coupled with a number of key energy sector developments, has dramatically changed the investment landscape. The current global financial crisis has created significant constraints from which the U.S. and European energy industries are not immune. Money is exceptionally tight, and will likely remain so for several years. As a result, overall momentum and public and private

[8] International Energy Agency, "Energy Technology Perspectives 2008: Scenarios and Strategies to 2050," Paris, available at http://www.iea.org/w/bookshop/add.aspx?id=330.

[9] Franklin Kramer, John Lyman, "Transatlantic Cooperation for Sustainable Energy Security: A report of the Global Dialogue between the European Union and the United States," The CSIS Press, 2009.

[10] Laura Diaz Anadon, Kelly Sims Gallagher, Matthew Bunn, and Charles Jones, "Tackling US Energy Challenges and Opportunities: Preliminary Policy Recommendations for Enhancing Energy Innovation in the United States," Cambridge, Mass: Energy Research, Development, Demonstration & Deployment Policy Project, Energy Technology Innovation Policy Group, Harvard University, February 2009.

investments in new research and development projects are slowing. This will dramatically inhibit the pace at which the U.S. and European energy industries can be transformed.

Although tentative signs of recovery from the global financial and economic crisis are gaining strength, policymakers around the world are still grappling with the effects of the crisis on the real economy. In the United States, unemployment is still historically high and credit is still constrained. The International Labor Organization predicts that employment levels in those countries with a high gross domestic product (GDP) per capita will not return to pre-crisis levels before 2013.[11] Furthermore, poor economic conditions are now recalibrating constituents' concerns. For instance, a majority of Americans recently told the Gallup Poll, for the first time in Gallup's twenty-five year history of asking the question, that economic growth should be given priority over environmental protection, even if the environment suffers to some extent.[12] A Pew Research Center survey of the public's priorities reports that global warming is now in last place, having dropped 10 percentage points, to 28%, from 2007.[13] Unless the measures to mitigate climate change, such as investments in energy innovation, are tied to and determined by more pragmatic approaches, such as energy security and job creation, economic conditions will be shaping public concerns.

While billions were spent on *green* investments on both sides of the Atlantic as part of the stimulus packages created in response to the economic crisis, domestic short-term efforts will not be enough to generate long term technological innovation.[14] The need to rethink the prevailing paradigm for economic growth in the wake of the recent financial crisis has presented an opportunity to forge policies

---

[11] International Labor Organization, *World of Work Report 2009: The Global Jobs Crisis and Beyond* (Geneva, 2009).

[12] Frank Newport, "Americans: Economy Takes Precedence over Environment," Gallup Poll, March 19, 2009.

[13] Pew Research Center for the People and the Press, "Public's Priorities for 2010: Economy, Jobs, Terrorism," January 25, 2010.

[14] Nick Robins and Robert Clover, "A Climate for Recovery: The Colour of Stimulus Goes Green," HSBC Global Research, February 16, 2009.

that at once meet urgent economic and social needs while finding a new, low-carbon path to prosperity and growth.

Using current technologies, the investments needed simply to meet projected increases in energy demand by 2035 is a staggering $38 trillion.[15] Furthermore, without significant innovation in the underlying technologies, models of a cost-effective global climate mitigation policy[16] suggest that the cost of greenhouse gas mitigation through 2050 would require additional trillions or tens of trillions of dollars.[17] The enormous magnitude of the costs imposed on the global economy creates an immediate imperative for innovation in the energy sector.

Any investment, even more so the high level of investment required by the energy industry, requires a clear and predictable regulatory environment. Currently, the energy industry operates in an environment that lacks a definitive clean energy standard, and this policy uncertainty increases risks for clean energy projects.[18] Price signals on carbon have the ability to release a wave of innovation and investment in green energy that will be needed to make clean technology cost-competitive and sustainable well into the future.[19] Yet, the last comprehensive legislation in the U.S., The American Power Act, is now history and there is no other viable initiative for the foreseeable future.[20] While the EU has some of the highest environment standards in the world, it is still confronted with similar roadblocks in pro-

---

[15]International Energy Agency, "Cumulative investment in energy infrastructure, 2011-2035," Available at http://www.iea.org/weo/Files/2011_EBC_Ministerial_Press.pdf.

[16]For atmospheric stabilization targets in the range of 450–550 parts per million CO_2.

[17]K. Gillingham, R.G. Newell, W.A. Pizer, "Modeling endogenous technological change for climate policy analysis," *Energy Economics*, vol. 30 no. 6, November, 2008, pp. 2734-2753.

[18]Kira R. Fabrizio, "Investments Under Regulatory Uncertainty: Evidence from Renewable Energy Generation,"Fuqua School of Business, Available at http://faculty.fuqua.duke.edu/bio/fabrizio/papers/pdfs/InvestUnderUncertainty_Fabrizio.pdf.

[19]World Economic Forum, "The Green Investing: Towards a Clean Energy Infrastructure," January 2009, Available at https://members.weforum.org/pdf/climate/Green.pdf.

[20]The political divide is so great on these issues that there is significant opposition even to the 2011 EPA initiative on emissions standards for power plants, designed in the interest of something more politically palatable such as public health.

viding clear market signals. The recent European debate over raising the greenhouse gas emissions reductions target to 30% of 1990 levels by 2020, from the previously established 20%[21], reflects the same policy uncertainty that discourages the much-needed investments in research, development and deployment on clean energy technologies.

The current global economic crisis, the shifting perceptions of the threat of climate change and the lack of a clear regulatory environment act as constraining factors in the development of energy innovation. Thus, it is even more critical for resources to be applied to the most promising technologies and that the most efficient policies, programs and regulations are effectively implemented. Extensive transatlantic cooperation is required to create vigorous distribution of ideas and experiences if financial and technical knowledge are to be used most effectively.[22]

The Role of Transatlantic Cooperation in Energy Innovation

New technologies need to be developed to increase supply options and to improve efficiency of demand as well as of production. Solutions to energy needs require both the U.S. and the EU to deal with many of the same issues:

First, planning and executing a plan to address the resources, technologies, and human capabilities needed to build the required infrastructure entails very long lead times. Additionally, the need exists to develop the institutional, structural and professional capabilities that will enable the continued acceleration of technological developments and will result in long-term sustainable growth.[23] This process will usu-

[21]Nigel Purvis, "Weathering the Transatlantic Climate Policy Recession," German Marshall Fund of the United States, March 2011, Available at http://www.gmfus.org/publications/publication_view?publication.id=1610.

[22]Franklin Kramer, John Lyman, "Transatlantic Cooperation for Sustainable Energy Security: A report of the Global Dialogue between the European Union and the United States," The CSIS Press, 2009.

[23]Richard G. Newell, "A US Innovation Strategy for Climate Change Mitigation,", A Hamilton Project Discussion Paper, The Brookings Institution, December 2008, Available at http://www.brookings.edu/papers/2008/12_climate_change_newell.aspx.

ally transcend several changes in governments and administrations over many years, and therefore must involve long-term commitments.[24]

Second, matching the availability of energy supplies with demand while reducing emissions will involve cultural and lifestyle changes that may be very difficult to achieve in the political realm. Thus, the rationale for such changes and the benefits of new technology options must be compelling and well understood by the governments and populations affected.[25]

Third, the factors that constrain the development of energy innovation call for the following major issues to be addressed: insufficient financial resources; inadequate institutional arrangements;[26] lack of sector coordination and lack of long term political commitment; insufficient information and communication;[27] and inadequate human resources.

While the responses of individual countries to these challenges are vital and cannot afford to be curbed, without a high degree of transatlantic cooperation these objectives and the policies designed to meet them will not be sufficiently resilient to weather volatile economic conditions and changing political currents.

Furthermore, the borders of an energy innovation system do not coincide with national borders. Countries can exploit the new energy technologies and knowledge originating from energy R&D that other countries have developed via international trade, multinationals and

[24]Charles W. Wessner, "New Vistas in Transatlantic Science and Technology Cooperation," Board on Science, Technology, and Economic Policy, National Research Council, 1999, Available at http://www.nap.edu/catalog/9455.html.

[25]Rebecca M. Henderson, Richard G. Newell, ed., *Accelerating Energy Innovation: Insights from Multiple Sectors*, University of Chicago Press, pg. 1–23.

[26]Michael Mehling, Aaron Best, Dominic Marcellino, Michael Perry, Katharina Umpfenbach, "Transforming Economies through Green Investment: Needs, Progress and Policies," German Marshall Fund of the Unites States, January 2010, Available at http://www.gmfus.org/galleries/ct_publication_attachments/GMFEcologicTransformingEconomies.pdf.

[27]Bracken Hendricks, Sean Pool, Lisbeth Kaufman, "Low-carbon Innovation: A Uniquely American Strategy for Industrial Renewal," Center for American Progress, May 2011, Pg. 33-34, available at http://www.americanprogress.org/issues/2011/05/pdf/gcn_low_carbon.pdf.

international knowledge spillovers.[28] The potential for a free rider scenario to occur creates incentives for pooling knowledge, technological and financial resources across borders and strengthens the case for international cooperation in energy R&D.[29]

Since much of the needed technology is not yet available, sustained and substantial research, development and demonstration will be required to bring it to market. The capital requirements will be significantly larger than what private industry can currently support and there is a definite role for government support. However, in an environment where appropriate policies are lacking, energy producers and users will not fully adopt the new energy technologies, because by using existing technologies they can avoid bearing the entire environmental costs they create.[30] This well-known problem is caused by the public good nature of the benefits that these technologies generally engender. To address this market failure a portfolio of policies is needed that combines government support, technological standards, financial penalties and awards with market based instruments. In addition, the benefit for sustained research, development and demonstration will be the design of more cost effective solutions and the avoidance of the inherent costs associated with poor policy decisions.[31]

While it is important to create domestic regulations and incentives for innovation, without cooperation there is a danger that the transatlantic community may create unnecessary differences in implementation that will stifle trade and reduce the potential to achieve economies of scale.[32] Moreover, the lessons learned from the experi-

---

[28]Valentina Bosetti, Carlo Carraro, Emanuele Massetti, Massimo Tavoni, "International energy R&D spillovers and the economics of greenhouse gas atmospheric stabilization," *Energy Economics*, 30, 2008, 2912-2929.

[29]David Popp, "Innovation and Climate Policy," NBER Working Paper No. 15673, January 2010.

[30]Rebecca M. Henderson, Richard G. Newell, ed., *Accelerating Energy Innovation: Insights from Multiple Sectors*, University of Chicago Press, pp. 1–23.

[31]Richard Lawson, John Lyman, Mihaela Carstei, "A Shared Vision for Energy and Climate Change: Establishing a Common Transatlantic Agenda," The Atlantic Council of the United States, 2010.

[32]Franklin Kramer, John Lyman, "Transatlantic Cooperation for Sustainable Energy Security: A report of the Global Dialogue between the European Union and the United States," The CSIS Press, 2009.

ence of U.S. state regulators and European national regulators in regulating similar issues can provide opportunities to design more effective policies and incentives. Transatlantic cooperation will also make it easier to obtain international consensus on establishing technical standards and benchmarking of industries.[33]

Cross-border trade is a significant channel for knowledge flows and R&D spillovers, and international investment both, spurs the development and introduction of new technologies and business methods, and provides for healthy competition that fosters innovation. As the U.S. and EU have the most advanced economic relationship on the planet and trade between the two markets is key to businesses on both sides of the Atlantic, encouraging transatlantic investment is a crucial part of fighting the effects of global economic crisis and increasing potential economic growth with new and innovative technology.[34] The deep economic integration between the U.S. and EU can also accelerate the return to economic growth in both areas. Furthermore, the large multinational companies that expand operations to the other side of the Atlantic include some of the most innovative and technologically forward-looking companies in both the U.S. and Europe. Transatlantic cooperation is thus essential in creating an environment conducive to cross-border investments that lead to innovation as well as relieves some of the current global economic pressures.[35] Through cooperation the transatlantic community can share the costs, risks, and resources required for basic scientific research and long-term propositions; reduce the costs of emerging technologies through accelerated learning; share costs for research and development on technologies that are expected to provide public benefits; promote the development

[33]Richard Lawson, John Lyman, Mihaela Carstei, "A Shared Vision for Energy and Climate Change: Establishing a Common Transatlantic Agenda," The Atlantic Council of the United States, 2010.

[34]Transatlantic Business Dialogue, "Accelerating the Transatlantic Innovation Economy: 10 Innovation Policy Principles & Recommendations to Strengthen Collaboration across the Atlantic", October 2010. http://www.tabd.com/index.php?option=com_content&view=section&layout=blog&id=8&Itemid=11.

[35]Ibid.

of innovative capabilities at home and abroad to promote long-term competitiveness; and develop mutually acceptable solutions to common problems.[36]

The importance of transatlantic cooperation is difficult to dismiss. The urgency and scale of needs creates an imperative for the U.S. and EU jointly to develop solutions to the current and future technological, financial and social challenges. The cost of developing new technologies places a tremendous strain on any single economy and innovation in the energy sector has wide ramifications throughout the economy, impacting trade and economic growth. Nonetheless, without significant technological progress to address climate change, the UNEP estimates that global economic cost could amount to 5-10% of global domestic product.[37] Accelerating innovation and technology adoption in energy is thus crucial to meeting greenhouse gas mitigation goals. Thus, the objectives of achieving secure and affordable energy that fosters economic prosperity will not be attainable in the absence of transatlantic cooperation.

Policy Recommendations for Transatlantic Cooperation

The transatlantic community is uniquely positioned to develop the technology, financing, and legislative and regulatory developments necessary to advance energy security. Further, there is growing recognition throughout the transatlantic community that the energy sector needs to be radically transformed in order to achieve energy security and to reduce the impact of climate change while ensuring economic prosperity. Thus, pursuing an interrelated set of strategic objectives will ultimately require similar structural changes in the way energy is

[36]Laura Diaz Anadon, Kelly Sims Gallagher, Matthew Bunn, and Charles Jones. *Tackling US Energy Challenges and Opportunities: Preliminary Policy Recommendations for Enhancing Energy Innovation in the United States.* Cambridge, Mass: Energy Research, Development, Demonstration & Deployment Policy Project, Energy Technology Innovation Policy Group, Harvard University, February 2009.

[37]Edward B. Barbier, "Rethinking the Economic Recovery: A Global Green New Deal," Report prepared for Economics and Trade Branch, Division of Technology, Industry, and Economics, United Nations Environment Program, April 2009, available at http://www.unep.org/greeneconomy/GlobalGreenNewDeal/tabid/1371/language/en-US/Default.aspx.

supplied and used throughout the transatlantic community. Moreover, the transformation of the energy sector will entail considerable analysis of new policies, regulations and enforcement mechanisms, as well as the deployment of many new technologies that are promising, but not yet proven. The foundation for transatlantic energy security will be the establishment of common, compatible and complementary strategies that allow for enhanced innovation.[38]

Existing cooperation on subjects ranging from climate change to R&D cooperation on specific technologies is essential but not sufficient. The range of organizations focusing on advancing energy innovation is encouraging. Institutions such as the EU-US Energy Council, the Transatlantic Business Dialogue, and the U.S. Chamber of Commerce as well as other international organizations, private industries, universities, and nongovernmental organizations are developing initiatives that have the potential to significantly strengthen the understanding of issues, broaden the political will, and provide the transatlantic community with useful information related to the feasibility and implementation issues involved in transforming the energy economy. However, these initiatives are not sufficient by themselves to drive nations on both sides of the Atlantic to undertake the radical transformations deemed necessary to address climate and energy security concerns.[39]

The needed changes in the energy sector will potentially result in a complete reorganization of our society. Thus, the framework for transatlantic cooperation needs a broader focus on innovation as it permeates throughout the different economic sectors and on areas of cooperation which will charge the bureaucracy on both sides of the Atlantic towards clearly defined goals.[40]

---

[38]Franklin Kramer, John Lyman, "Transatlantic Cooperation for Sustainable Energy Security: A report of the Global Dialogue between the European Union and the United States," The CSIS Press, 2009.

[39]Nigel Purvis, "Weathering the Transatlantic Climate Policy Recession," German Marshall Fund of the United States, March 2011, Available at http://www.gmfus.org/publications/publication_view?publication.id=1610.

[40]Richard Lawson, John Lyman, Mihaela Carstei, "A Shared Vision for Energy and Climate Change: Establishing a Common Transatlantic Agenda," The Atlantic Council of the United States, 2010.

Policies aimed at accelerating deployment of energy technologies are often decoupled from key energy policy goals and poorly linked with policies for research, development and deployment.[41] The U.S. and the EU should target and coordinate incentives for large-scale deployment of energy technologies as key elements of a comprehensive energy innovation strategy. Furthermore, policy makers on both sides of the Atlantic should take advantage of the increasingly collaborative and cross border nature of innovation to create a foundation for transatlantic harmonization of innovation policies on a broader scale. While greater focus is needed on both policy and technical challenges that new technologies will face as they enter the market, many of the challenges to innovation result from existing competition and entrenched political interests.[42] For example, one of the most effective ways to encourage development and deployment of new technologies is to set a price on carbon; yet, the political will to pass enabling regulation is nonexistent in the U.S. for the foreseeable future.

Nonetheless, while many issues between the U.S. and EU are highly technical and are best dealt with at the working level, high-level working agendas should address broad areas for enhanced cooperation in which clear goals for action are established. They should:

Develop and support transatlantic public research initiatives that are tightly linked to the private sector and acknowledge and respond to non-technical challenges to innovation. Policy makers on both sides of the Atlantic should create effective regulatory frameworks and long-term incentives for public-private partnerships. These would necessarily include: developing new models for the funding, commercialization and deployment of new innovative advances; funding for precompetitive research that can aid in bringing technologies their commercial stage;[43] building human capital and knowledge stocks by increasing

---

[41]Charles Weiss, William B. Bonvillian, "Structuring an Energy Technology Revolution," The MIT Press, pg. 1-11, Available at http://mitpress.mit.edu/catalog/item/default.asp?ttype=2&tid=11808.

[42]Clean Air Task Force, "Four Policy Principles for Energy Innovation & Climate Change: A Synthesis," June 2010, Available at http://www.catf.us/resources/publications/view/125.

[43]Transatlantic Business Dialogue, "Accelerating the Transatlantic Innovation Economy: 10 Innovation Policy Principles & Recommendations to Strengthen Collaboration across the Atlantic", October 2010. http://www.tabd.com/index.php?option=com_content&view=section&layout=blog&id=8&Itemid=11.

the availability of transatlantic scholarships and other financial support for the science fields;[44] support the use of performance based technology regulation and standards; promote the creation of a robust information technology infrastructure; and encourage investment in innovative technologies. This mix would allow for public sector investments to fully leverage private capital at scale as well as for private resources to capitalize on public expenditures.

Create a competitive market for transatlantic energy innovation. A key policy mechanism to encourage and promote innovation in the development and deployment of energy technologies is the removal of trade and investment barriers. A transatlantic commitment to promote open investment policies, avoid new trade restrictions, and aim to eliminate existing barriers is an acknowledgement that innovation is fundamental to our shared prosperity and creates an attractive environment for much needed capital flows.[45]

Create common transatlantic intellectual property and standards policies that promote vigorous competition. Policies to reward and protect innovation are equally important. The transatlantic community should aim to develop complementary competition and property laws that encourage innovation while enhancing economic efficiency and consumer welfare.[46] Furthermore, policy makers should establish a framework that allows for the rights to intellectual property to be shared appropriately and emphasizes the commitment of governments on both sides of the Atlantic to ensuring consistent and effective enforcement.[47]

---

[44]Richard Lawson, John Lyman, Mihaela Carstei, "A Shared Vision for Energy and Climate Change: Establishing a Common Transatlantic Agenda," The Atlantic Council of the United States, 2010.

[45]Transatlantic Business Dialogue, "Accelerating the Transatlantic Innovation Economy: 10 Innovation Policy Principles & Recommendations to Strengthen Collaboration across the Atlantic", October 2010. http://www.tabd.com/index.php?option=com_content&view=section&layout=blog&id=8&Itemid=11.

[46]Franklin Kramer, John Lyman, "Transatlantic Cooperation for Sustainable Energy Security: A report of the Global Dialogue between the European Union and the United States," The CSIS Press, 2009.

[47]Transatlantic Business Dialogue, "Accelerating the Transatlantic Innovation Economy: 10 Innovation Policy Principles & Recommendations to Strengthen Collaboration across the Atlantic", October 2010. http://www.tabd.com/index.php?option=com_content&view=section&layout=blog&id=8&Itemid=11.

The energy and economic challenges facing the transatlantic community, as well as the world, are daunting. More effective transatlantic cooperation in energy technologies, to reduce the costs and risks of energy innovation and to increase the pace of cost reductions through expanded learning and deployment, can help ensure a future that meets the goals of energy security, environmental responsibility and economic prosperity. Furthermore, by creating a comprehensive transatlantic strategy for and investment in energy innovation, and by identifying new approaches and policies for bringing new technologies to market, the transatlantic alliance can meet these challenges and renew and restore innovation leadership within the global community.

Chapter Five

To Begin the World Anew: Smart Grids and the Need for a Comprehensive U.S. Energy Policy

Nikolas Foster

The United States is in the midst of a monumental transformation of its electric power system. Advances in information and communication technologies and grid measurement and control devices have initiated the transition toward a more resilient, sustainable and efficient future power grid. Deployment of these technologies is being driven by policies encouraging the shift to a greener and more secure grid, incorporating clean and low carbon energy as well as rising consumer demand for smarter ways to use existing resources.

While the passage of the Energy Independence and Security Act (EISA) in 2007 and the 2009 American Recovery and Reinvestment Act (ARRA) accelerated U.S. investment in a "smart grid" sufficiently flexible to accommodate technological and societal changes associated with transition in the nature of the electric generating fleet, the U.S. is not alone in its grid transformation. Countries across the globe have and continue to invest billions in their electricity transmission and distribution systems in an effort to adjust to the challenges of rising demand, constrained resources and growing environmental consciousness. In its effort to modernize its grid and create a large-capacity interconnected transmission backbone, China spent more than $7.3 billion on smart grid stimulus investments in 2010.[1] Efforts under way in the European Union (EU) also warrant a closer look. In 2010, Italy alone had about three times the amount of smart meters installed

[1] SmartGridnews.com (2010, January 27), Smart Grid Snapshot: China Tops Stimulus-Funding, SmartGridNews.com, available at: http://www.smartgridnews.com/artman/publish/Stimulus_News_Digest_Products/Smart-Grid-snapshot-China-Tops-Stimulus-Funding-1827.html, accessed August 2011.

compared to the U.S.[2] EU wide, there are more than 45 million smart meters installed, keeping the continent on track to fulfilling one of the European Commission's smart grid deployment goals—structured smart grid deployment milestones the United States does not have.[3] Even so, the U.S. has recently established 99 smart grid investment grant programs across the nation, spending $3.4 billion and unleashing more than that in private sector investments.[4]

This chapter explains the differing motivations behind smart grid deployment in the United States and Europe. It will also aim to examine the progress made in the United States, give an overview of recent smart grid-related research and outline policy recommendations for further deployment. Lastly, it suggests areas of transatlantic cooperation that would help both the United States and countries in the EU accelerate smart grid deployment.

What is the Smart Grid?

The smart grid is a network of transmission lines, interactive equipment, and new technologies working together to help consumers and producers of energy save economic and environmental resources. This digitization of the power system delivers sensing, communications and controls across the system from transmission to distribution to the consumer, delivering a two-way flow of energy and communications that provide real-time system transparency to operators and consumers alike.

---

[2] Zhang, Zhen. "Smart Grid in America and Europe—Similar desires, different approaches, Part II" *Public Utilities Fortnightly*, February 2011, p. 33.

[3] Giordano, Vincenzo, et al. "Smart Grid projects in Europe: lessons learned and current developments," *European Commission Joint Research Centre Institute for Energy*, p. 13.

[4] "Follow the Money: Stimulus Funding Begins to Flow Into Smart Grid Sector," available at: http://www.kema.com/services/ges/smart-grid/AI/follow-the-money-stimulus-funding-begins-to-flow-into-smart-grid-section.aspx.

Smart Grids of Necessity, Smart Grids of Choice

Both the United States and the EU are moving in the direction of the smart grid, yet at a varying pace and for often different reasons. One similar characteristic of this movement is that progress in the United States and in Europe is not homogeneous, but differs internally depending on regulatory, policy, regional and cultural variables.

There are a multitude of stakeholders involved in generating, transmitting, distributing, and regulating electricity in the U.S. In 2007, there were 9,554 utilities across the nation, of which 1,934 were generating electricity.[5] Policy is made by a combination of federal and state entities. For example, the U.S. Department of Energy oversees technology research and development, and has a role in examining certain policy issues such as transmission siting. However, the Federal Energy Regulatory Commission (FERC) oversees regulatory questions relevant to wholesale power markets and interstate transmission, including grid reliability. The Environmental Protection Agency (EPA), meanwhile, administers the Clean Air Act, which impacts emissions-related requirements associated with electric generation. If the alphabet's soup of federal entities were not enough, it is in fact State boards and commissions—some appointed by Governors, others elected by popular ballot—that regulate retail sales and the distribution of electricity to customers.

Moreover, it is important to appreciate that the U.S. electricity sector is highly differentiated on the state level as well. Texas has a deregulated retail market for electricity, unlike any of its neighboring states, allowing most of its residents to select their retail electric provider from a list of utilities.[6] Restructuring of electricity markets is currently on hold in a number of jurisdictions in the aftermath of the western electricity crisis of 2000-2001, which was widely perceived as an economic disaster driven by a combination of bad actors, bad weather and poor market design in California.[7] Meantime, the

---

[5] U.S. Census Bureau, "2007 Economic Census." Available at www.census.gov/compendia/statab/2012/tables/12s0923.pdf, accessed October 2011.

[6] See http://www.powertochoose.org/.

[7] "Adressing the 2000-2001 Western Energy Crisis," available at http://www.ferc.gov/industries/electric/indus-act/wec/chron.asp.

debate about restructuring has gone completely dormant in a majority of states.[8]

In the EU, the situation is similar in terms of comparing member states' electricity sectors. Countries within the EU have differing electricity markets, ranging from the centrally organized and fully nationalized electricity market in France to the United Kingdom's competition-based system.[9] While the EU does not have a central regulatory authority equivalent to FERC, it does have political mechanisms that can direct changes to the continent's electricity sector through decisions by the European parliament and resulting directives from the European Commission. One of the recent milestones here is the "Third Energy Package," the tenets of which include an 80% penetration of smart meters in European households by 2020.[10] The EU has also tied smart grid to its general goal of unbundling transmission systems operation, building a liberalized internal market for electricity, incorporating renewable energy and reducing carbon emissions.[11] Article 3, section 11 in chapter 2 from the above mentioned legislation calls on member states to develop "innovative pricing formulas," and introduce "intelligent metering systems or smart grids."[12] Reading the plans and directives suggests that incorporating renewable energy and reducing the continent's carbon footprint are the EU's main drivers of advancing smart grid deployment. An April 2011 report from the European Commission calls the smart grid the "backbone of the future decarbonised power system," which will "enable the integration of vast amounts of both on-shore and off-shore renewable energy."[13] EU pol-

---

[8] "Status of Electricity Restructuring by State," Energy Information Administration, available at http://www.eia.gov/cneaf/electricity/page/restructuring/restructure_elect.html.

[9] Von Danwitz, Thomas, "Regulation and Liberalization of the European Electricity Market—A German View," *Energy Law Journal*, Volume 27, p. 423.

[10] Electricity Directive (2009/72/EC).

[11] Directive 2009/72/EC, available at http://eur-lex.europa.eu/LexUriServ/LexUriServ.do?uri=CELEX:32009L0072:EN:NOT.

[12] Supra 8.

[13] "Smart Grids: From Innovation To Deployment" Communication From The Commission To The European Parliament, The Council, The European Economic And Social Committee And The Committee Of The Regions, available at http://eur-lex.europa.eu/LexUriServ/LexUriServ.do?uri=COM:2011:0202:FIN:EN:PDF.

icy also states that the smart grid will "significantly reduce the environmental impact of the whole electricity supply system."[14] Along the same lines, the Strategic Deployment Document for Europe's Electricity Networks picked smart grid to become an essential enabler for the continents 20/20/20 goals, cutting emissions and increasing renewable energy by 2020.[15]

Beyond directives by the European Commission determining many of the smart grid's features, the EU's smart grid policy is guided by several plans and initiatives, including the SmartGrids European Technology Platform for the Electricity Networks of the Future (SmartGrids ETP), the European Electricity Grid Initiative (EEGI) Roadmap and Implementation Plan, the EU Framework Program, and the European Strategic Energy Technology Plan (SET Plan).[16] These plans and initiatives prioritize deployment strategies, outline specific milestones and determine funding levels for smart grid research.

But while it is arguably the case that environmental outcomes have animated EU policy in relation to the smart grid, U.S. entities—driven by public/private partnerships—significantly accelerated efforts in the areas of advanced technology development and deployment as a result of the Northeast/Midwest blackout of August 2003. The blackout left more than 50 million people in cities including New York City, Detroit, Cleveland and Toronto without electricity for up to four days, drove policymakers to adopt federal legislation putting in place mandatory reliability standards governing the electric utility industry, and reinvigorated research on technologies designed to provide enhanced visibility over grid systems.

In addition, peak demand, a spike in electricity consumption usually occurring in the afternoon and early evening hours of the winter and summer months, is challenging grid operators and has been another prime motivator for smart grid deployment. Over the last few decades,

---

[14]The SmartGrids European Technology Platform, available at http://www.smartgrids.eu/node/81.

[15] "Strategic Deployment Document for Europe's Electricity Networks," *European Technology PlatformSmartGrids*, p. 6.

[16]Zhang, Zhen. "Smart Grid in America and Europe—Similar desires, different approaches, Part I." *Public Utilities Fortnightly*, January 2011, p. 48.

end-use consumption of electricity has grown faster than that of petroleum or natural gas.[17] The residential sector has, driven by the growth of air-conditioners in housing from 68% in 1993 to 87% in 2009, as well as the proliferation of consumer electronics, outpaced both commercial and industrial users.[18]

For generators of electricity, peak demand translates into a vexing challenge: during most of the year, about 50% of the country's generation capacity is fully used. However, for 5% of the time, or about 400 hours every year, more than 90% of the capacity is used.[19] These hours are crunch times for grid operators, typically occurring during the hottest days of summer when customers turn on their air conditioners, televisions and other electronic appliances. Until now, utilities reaction has been to "build to peak;" building new peak generation plants and then firing them up in times of need. Back-up generation through peak generators is, however, the least economic- and most emissions-intensive generation; utilities pay hundreds of millions of dollars every year just to provide electricity during these few hours.

Informed by these events and developments, the EISA of 2007 stated that the main purpose of a smart grid was to "maintain a reliable and secure electricity infrastructure that can meet future demand growth."[20] Delaying or removing the need for expensive generation capacity expansion by addressing the problem of peak demand is thus a prime policy objective.

Comparing the motivations behind U.S. and EU smart grid deployment thus illustrates a distinct characteristic. While the nature of the U.S. power grid has led to a policy focus on reliability, interoperability and advanced research and development—with deployment

---

[17]Joskow, Paul; "Challenges for Creating a Comprehensive National Electricity Policy," Speech at Technology Policy Institute, September 26, 2008.

[18]2009 Residential Energy Consumption Survey 2009, Energy Information Administration, available at http://www.eia.gov/consumption/residential/reports/air_conditioning09.cfm and http://www.eia.gov/consumption/residential/reports/electronics.cfm.

[19]Lightner,Eric,Widergren, Steven."An Orderly Transition to a TransformedElectricity System." *IEEE Transactions On Smart Grid*, Vol. 1, No. 1, June 2010, p. 4.

[20]"Energy Independence and Security Act of 2007, Title XIII, Section 1301, available at http://energy.senate.gov/public/_files/getdoc1.pdf.

driven by Federal matching funds for utility smart grid investments—EU policymakers have also deemed smart grid as an enabler for the liberalization of electricity markets and a means to address climate change.[21]

Benefits of the Smart Grid

While reliability and peak demand issues have been primary, initial drivers of smart grid deployment in the United States, there is increasing recognition that building additional intelligence into grid infrastructure can help achieve a number of additional energy policy goals, including reducing U.S. dependence on foreign oil, lowering the emissions profile of the electric generating fleet, decarbonizing transportation and enhancing the resilience of infrastructure critical to economic and national security.

Renewables Integration

In the United States, 24 states plus the District of Columbia currently have a Renewable Portfolio Standard (RPS).[22] These standards mandate energy providers selling energy within a state to produce or purchase a specific amount of renewable energy by a determined date. Together with State Renewable Energy credits and Federal Production Tax Credits (which, depending on the form of energy, are bound to expire by December 2012 and 2013[23]), RPSs have contributed to a rapid growth of renewable energy across the nation. In 2009, more than 60% of total renewable electric generation additions came from wind power alone.[24]

While the growth in wind power and other renewables is supporting President Obama's goal of generating 80% of electricity from

[21]Supra 19, Section 1304-06.

[22]Five other states, North Dakota, South Dakota, Utah, Virginia, and Vermont, have nonbinding goals for adoption of renewable energy instead of an RPS. http://apps1.eere.energy.gov/states/maps/renewable_portfolio_states.cfm.

[23]See American Recovery and Reinvestment Act of 2009, H.R. 1 (Div. B, Section 1101 & 1102) available at http://thomas.loc.gov/home/h1/Recovery_Bill_Div_B.pdf .

[24]Electric Power Industry 2009: Year in Review, Energy Information Agency, (Washington, D.C., November 2010, updated January 2011 & April 2011).

clean sources by 2035, the intermittent character of such generation is creating headaches for grid operators worried about fluctuations in electricity generation. Here, smart grid technologies such as near real-time sensor and monitoring technologies will mitigate many of the existing uncertainties by combining energy storage, better forecasting and access to resources generated outside a control area.[25]

Electric Mobility

Recent research in electricity fueled transportation points toward additional benefits for grid operators and customers coming from battery storage in plug-in electric vehicles used as balancing tools for growing amounts of renewable energy. Scientists at my laboratory examined a decentralized control scheme called Grid Friendly Charging, which varies the charging rates for electric vehicles (EV) depending on the state of the grid. At night, when abundant wind power is available, the Smart Charger would allow EVs to absorb surplus energy and recharge their batteries, using the additional renewable power. During peak load times, this technology would postpone EV charging, and use stored electricity from EV batteries to strengthen the grid.[26]Grid Friendly Charging will support electricity operators' goals of having a more predictable load by balancing demand for electricity. In combination with real time pricing, which regulators across the United States are now testing in select regions, smart charging would bolster the business case for electric vehicles and the smart grid itself by allowing customers to charge their vehicles at the most economic rate and, where possible, even sell surplus electricity back to the grid. Lastly, substituting more transportation fuels away from fossil sources to domestically produced renewable energy would address a key policy goal- reducing their dependency on foreign fuel and curbing the transportation sector's carbon footprint.

[25]Imhoff, Carl. "Grid 2050—Shaping Grid Transformation," *Public Utilities Fortnightly*, August 2011, p. 32.

[26]Tuffner, F. K., Kintner-Meyer, M, "Using Electric Vehicles to Mitigate Imbalance Requirements Associated with an Increased Penetration of Wind Generation," IEEE Power and Energy Society General Meeting 2011, p .7.

CO_2

In addition to supporting the integration of renewable energy and clean transportation, a fully deployed smart grid by itself could also reduce CO_2. During the next twenty years, smart grid could lower the U.S. carbon emissions and energy use of the electric sector by up to 12%.[27] By combining smart grid technologies, such as consumer information and feedback systems, conservation voltage reduction, and building diagnostics, the grid of the future would create significant efficiency gains and demand reductions. In emissions terms, these savings would result in the United States preventing the equivalent of 442 million metric tons, or 66 typical coal power plants' worth, of carbon emissions from entering the atmosphere each year. In energy terms, these 66 power plants produce the equivalent amount of electricity needed to power 70 million of today's homes.

While the United States appears far from managing its carbon emissions by means of a carbon tax or installing a cap-and-trade system, its emissions-reducing character would make the smart grid an even more attractive investment driving CO2 reduction, if and when carbon management schemes were to be introduced.

Smart Grid—Strong Economy

Smart Grid will allow the United States to grow a domestic industry around energy provision, delivery and demand management. Recent private and public smart grid investments in the United States alone are estimated to directly create more than 280,000 jobs between 2009 and 2012, of which half are expected to be retained as permanent, high-skilled jobs.[28] These jobs will bring several economic benefits to communities across the United States. In West Virginia, for example, the job creation through deployed smart grid is estimated to provide annual benefits of $215 million to the U.S. economy.[29] As

[27]Pratt, R. et al. "The Smart Grid: An Estimation of the Energy and CO_2 Benefits," Pacific Northwest National Laboratory, January 2010, p. vi.

[28]The U.S. Smart Grid Revolution—Smart Grid Workforce Trends 2011, GridWise Alliance, 2011.

[29]NETL (National Energy Technology Laboratory) (2010), *West Virginia Smart Grid Implementation Plan—Roadmap Framework*, U.S. Department of Energy (DOE), prepared for GridWeek 2010, October 18 2010, Washington, DC, available at

businesses in smart grid-related manufacturing, services and software development are located in diverse locations from California to North Carolina to Texas, similar job-related economic benefits can be expected across the United States.[30]

The transition to smart grid will, however, increase the training and skill requirements of workers in the energy sector. This transition is paralleled in all sectors of the labor market, which demands new skill sets to manage the ever growing complexities of everyday work. In the energy sector, the rising reliance on information and communication technology, as well as advanced diagnostics and measuring devices, translates into a shift towards a highly trained workforce, able to design, install, maintain and service smart grid infrastructure and technology.

Federally Funded Smart Grid Investments in the United States—An Overview

To spur deployment of the smart grid, the U.S. has passed two important pieces of legislation in recent years. In 2007, Congress laid out a roadmap for grid modernization policy as part of the Energy Independence Act (EISA). In addition to charting general policy trajectory (including the aforementioned focus on reliability and protection of grid infrastructure) EISA's Section 1306 created the Smart Grid Investment Matching Grant Program, designed to provide up to 20 percent of the costs for utilities and other stakeholders that invest in deployment of certain smart grid-related equipment. Less than 18 months later, the ARRA of 2009 not only reaffirmed the political commitment to smart grid, but also augmented EISA's financial support by providing full matching grants for smart grid projects. About $4.5 billion in federally-funded awards to utilities, research organizations, private companies, manufacturers, cities and others spurred more than $8 billion of public-private smart grid investments. The 99 recipients receiving ARRA grants gave ongoing US smart grid deployments an

---

http://www.smartgrid.gov/sites/default/files/pdfs/10182010_gw_wv_sgip.pdf, accessed August 2011.

[30]Lowe, Marcy, et al. "U.S. Smart Grid—Finding new ways to cut carbon and create jobs," Center on Globalization, Governance & Competitiveness, Duke University, April 2011, p. 25.

essential boost in the midst of the 2008–2009 recession. Some highlights of these grants include:

- ARRA funded the deployment of 877 phasor measurement units (PMUs), expanding the prior nationwide network of 200 by more than 400 percent.[31]
- Federal grant awards for advanced metering infrastructure (AMI) deployments under ARRA total $812.6 million to date, with total project values reaching more than $2 billion.[32]
- ARRA includes a $2.4 billion program designed to establish 30 manufacturing facilities for electric vehicle batteries and components. This funding is in addition to the aforementioned $4.5 billion in awards made under ARRA.
- ARRA funded the Center for the Commercialization of Electric Technologies (CCET) Smart Grid Demonstration Project, a demonstration-scale micro grid project in Texas.

ARRA also continued funding for the Advanced Research Projects Agency-Energy (ARPA-E) to conduct smart grid R&D projects, and the Energy Information Administration (EIA) to collect data from the 14 ongoing smart grid demonstration projects. These demonstration projects are allowing both policy makers and electricity operators to lower costs and quantify the benefits of wider smart grid deployment. Indeed, the 16 ongoing demonstration projects are designed to test technologies in a systems context and help chart the business case for further smart grid deployments, for utilities, electricity providers and grid operators alike. Taking into account the regulatory and energy portfolio characteristics of different regions of the U.S., the demonstrations test everything from time of use pricing to interoperability.

---

[31]Overholt P. 2010. "North American SynchroPhasor Initiative (NASPI) and DOE's Smart Grid Investment Grants." Presented at the EEI Transmission, Distribution and Metering Conference. April 11-14, 2010, Arlington, Virginia. Accessed October 8, 2011 at http://www.eei.org/meetings/Meeting%20Documents/2010-04-TDM-Tuesday-4-Overholt-Philip.pdf.

[32]U.S. Department of Energy. 2010a. "Recovery Act Selections for Smart Grid Investment Grant Awards—By Category." U.S. Department of Energy, Washington, D.C. Accessed March 22, 2011 at http://www.oe.energy.gov/DocumentsandMedia/Combined_SGIG_Selections_Category-12-16-10.pdf (undated webpage).

For example, the largest, regional ongoing demonstration, the Pacific Northwest Smart Grid Demonstration Project, is occurring in a geographical area with a significant amount of renewable energy penetration. The Pacific Northwest does not, however, have any dynamic pricing for consumers available. Consequently, this demonstration is focusing on integrating renewable energy and developing a "transactive control" system, which will translate the grid's supply and demand of electricity into a signal that is received by responsive assets, such as smart water heaters and electric vehicles. Those assets will automatically react to and shift demand away from the grid during current peak times.[33]

Implications for Future Federal Energy Policy Deliberations

ARRA and preceding legislations allowed smart grid deployment in the United States to progress rapidly in a very short amount of time. The funds made available through EISA and ARRA allowed utilities, private energy services companies, research organizations, and manufacturers to test smart grid technology and inform their decisions how to proceed. While federal funding for smart grid R&D continues with approximately $30 million per year, the government is gradually reducing its involvement in financing large scale smart grid deployment across the nation. As the federal government is slowly stepping back, the question of what mechanisms might guide further investments and planning for smart grid deployment remain. This is especially important given the hopes and expectations pinned to the electric power system. Yet while smart grid holds the promise of reducing dependence on foreign fuels, decarbonizing transportation, building a more resilient grid and strengthening the domestic economy, the realization of these promises is by no means a given. The U.S. electricity sector in general is lacking a guiding policy that could drive the transformation towards the grid of the 21st century.

---

[33]Previous demonstrations using transactive control, such as the Olympic Peninsula Test Bed, achieved peak load reductions of up to 20%. See Hammerstrom, D, et. al, Pacific Northwest GridWise(tm) TestbedDemonstration Projects, Part I. Olympic Peninsula Project, Pacific Northwest National Laboratory, October 2007.

Stuck in a limbo of structural fragmentation, cautious approaches to restructuring and decades-old regulatory paradigms, the electricity sector would benefit from a national debate framing the needs and the policy goals for this transformation. As with other successful transformative changes to vital areas of national interest, a national policy and regulatory reform is needed. Be it the natural gas industry or the telecommunications sector, successful changes in these areas historically have been driven by federal initiatives.[34] But the electricity sector is, in the words of Paul Joskow, former dean of the MIT Sloan School of Business, "the last reform holdout."[35]

What could these reforms look like? Just as smart grid is blurring the lines between generation, transmission, distribution and end use, a new regulatory framework might be informed by emerging technology capabilities that blur existing jurisdictional lines. Surely, there will be no "one size fits all" solution to the highly diverse grid of the Unites States. Yet as the smart grid is changing the grid landscape, regulatory paradigms need to adapt accordingly to capture achievable benefits.

Looking at four sectors of potential regulatory involvement should illustrate areas where regulatory changes could bolster smart grid deployment and secure several important policy goals. Developing a holistic plan addressing electricity regulation jurisdictions, transmission lines, spreading the benefits and costs of deployment, as well as developing tools to maximize the benefits for the labor market resulting from grid evolution is essential to fulfill smart grid's promises. The following four examples are by no means exclusive to successful regulatory involvement. Yet they illustrate the positive impact regulation could have in ensuring successful smart grid deployment.

Expanding Control Areas

To square the circle of providing an abundant supply of reliable energy and integrating vast amounts of renewable and intermittent energy, control areas need to consolidate or at least vastly increase their coordination. Without consolidation or coordination, using the

[34]Supra 14, p. 4-6.

[35]Supra 14, p. 7.

wider area's generation diversity and access to existing assets in surrounding control areas will not be possible and thus hinder successful and efficient deployment of renewable energy. One study analyzing the integration of renewable energy into regional power grids has concluded that there is a need to "substantially increase balancing area cooperation or consolidation," and, "enable coordinated commitment and economic dispatch of generation over wider regions."[36] Another study focusing on wind energy alone states that "large operating areas—in terms of load, generating units, and geography—combined with adequate transmission, are the most effective measures for managing wind generation."

Apart from supporting the integration of renewable energy, expanding operating areas brings along further benefits for the electricity sector, such as saving money. One previous study concluded that "operating separately and locally, individual [balancing authorities] would have to purchase more expensive balancing reserves to accommodate the variability and uncertainty from high penetration of [variable generation] in the future."[37] There are other examples of existing and potential savings resulting from the consolidation of control areas. The PJM Interconnection, a regional transmission organization that coordinates the movement of wholesale electricity in 13 states and the District of Columbia, is one example where consolidation between control areas has led to efficiency gains, cost savings and growth in power flows.[38] Upcoming studies by Pacific Northwest National Laboratory and the National Renewable Energy Laboratory calculate significant economic savings, exceeding $250 million, resulting from greater coordination and consolidation.[39]

Getting large amounts of renewable energy online also will require expanding the existing transmission grid significantly. Research shows

---

[36]"Western Wind and Solar Integration Study," GE Energy, May 2010.

[37]Y. V. Makarov, N. Zhou, P. Etingov, N. Samaan, J. Ma, R. Diao, and R.T. Guttromson "Analyzing of Balancing Authorities Cooperation Methods with High Variable Generation Penetration," *Proc. 9th International Workshop on Large-Scale Integration of Wind Power into Power Systems as well as on Transmission Networks for Offshore Wind Power Plants*, Québec City, Québec, Canada, October 18-19, 2010.

[38]Supra 14, p. 11.

[39]Supra 26, p. 33.

that without enhancing transmission lines, "substantial curtailment (shutting down) of wind generation would be required."[40] Clearly, this scenario would not be desirable given the capital intensive nature of renewable investments.

The 2011 announcement by the White House to fast track seven transmission projects in both the western and eastern interconnection is a great step forward.[41] This measure will both support the electrical grid's reliability and ability to integrate renewable energy, while it is also setting an example for fast tracking the procedural steps of coordinating with various grid stakeholders on statutory reviews, permits, and regulation in general.

Beyond Smart Meters—Grid Operators' Smart Grid Planning

In an emerging age of constrained resources, new environmental benchmarks and rapidly changing grid technology, grid operators are reexamining their approach to electricity delivery. In a smarter grid, the general notion of finding supply-side solutions for growing demand is now balanced with consideration of mechanisms that can incentivize demand side changes, such as transactive control enabling smart appliances at home or grid friendly charging for electric vehicles. With this new operational strategy comes the potential for a new culture of business and new opportunities for revenue on the demand side.

Instead of creating incentives that solely reward cost efficiency by means of reducing operating costs, regulation should motivate operators to embrace the next generation of smart grid technology. A good example here is the wide-scale deployment of advanced metering infrastructure (AMI). Utilities across the United States have modernized their operational area by installing smart meters. While AMI is an important technology, it is not an end in itself. Utilities see smart meters as a tool to reduce operating costs by reducing meter readings and non-technical energy losses. Yet smart meters are also key enablers once combined with other smart grid technologies, such as

---

[40]Eastern Wind Integration and Transmission Study, EnerNex Corporation, Revised Feb. 2011, p. 27

[41]Interagency Rapid Response Team for Transmission, The White House Council on Environmental Quality, available at http://www.whitehouse.gov/administration/eop/ceq/initiatives/interagency-rapid-response-team-for-transmission.

demand response applications. To help move the smart grid to the next level, regulators should work with utilities to identify how to leverage investments for multiple consumer value streams that are likely to emerge over the next decade. This way, regulators could ensure that smart grid technology will benefit consumers and utilities alike for a range of potential benefits long into the future. Additionally, sensible regulation in this area can bolster the ongoing changes in utilities' operational culture. These regulatory innovations for smart grid must look across the entire system, from transmission (today's PMU upgrades) to distribution (distribution automation) and the consumer interface (AMI, HEMS, smart charging) and consider new value streams that accrue from leveraging information across these traditional boundaries! Another example of a cultural change for operators is the new role utilities are playing at the consumer level. Previously, a utility's role ended at the front door or the electric meter. Now, with demand response through smart water heaters, clothes dryers and thermostats coming from within the house, utilities need to address the new responsibilities coming from their expanded reach. Regulators should encourage utilities to be creative with their consumer incentive programs to provide more customer-oriented services, beyond just energy delivery. Discussions focusing on alternative revenue streams through efficiency gains and quality of service instead of volumes delivered are helpful here.

Alternatively, regulators could incentivize utilities to use outside contractors to manage some of these new business areas. Just as U.S. utilities have started outsourcing ancillary services, such as tree cutting, to contractors, outside vendors can manage the in home activities for utilities which in this case stay focused on their core business. Either way, regulation has a role to play to help utilities in this changing environment.

As the integration of renewable energy is a key policy goal of the Obama Administration and the majority of U.S. states, regulators should examine the role utilities can play to support the transition to clean and homegrown energy. Yet there are still utilities across the U.S. that do not offer their customers options to select renewable energy as a source for their electricity. Additionally, too often, consumers are not even aware of the renewable choices they have. A recent survey of 10,000 energy consumers across 15 countries found that more than half of the respondents did not know whether their energy provider had a green energy program or not.[42] Without these green options available or known, the goal of increasing the amount of renewable energy becomes harder to obtain.

Utilities provide the most immediate, interactive opportunity with customers. They are often the first and only stop for consumers to receive information about energy use, clean and green energy options and general knowledge. Utilities also build and manage the last mile in energy delivery infrastructure and for that reason they harbor the key to unlocking effective policy. It is important for regulators to establish goals that highlight a easy and clear path for all utilities offering customers renewable energy purchasing options, and offering utilities ideas for informing their customers about these options.

Keeping the Customer Satisfied—Spreading Smart Grid Benefits to Rate Payers

Grid operators can also be a key leveraging tool to get one of the smart grids most important assets involved: the well-informed customer. Customers' participation is crucial to realize several of the demand response strategies that will help fulfill many of the smart grid's efficiency promises. Recent surveys explain that a well-informed customer was more likely to have a positive opinion of local smart grid deployment programs underway or proposed[43]—a fact vital for operators confronted with opposition to, for example, smart meters in California or new transmission lines in Germany.[44] Informed consumers furthermore equate smart grid deployment with energy benefits for their families and are more likely to change their energy usage patterns to meet specific goals. Yet customers are still not treated as a part

[42]IBM 2011 Global Utility Consumer Survey.

[43]Supra 43.

[44]Barringer, Felicity. "New Electricity Meters Stir Fears," *The New York Times*, January 30, 2011, available at http://www.nytimes.com/2011/01/31/science/earth/31meters.html?pagewanted=all, and Diehl, Joerg, "Die Trassen-Brecher, Spiegel Online, July 18, 2011. Available at http://www.spiegel.de/wirtschaft/unternehmen/0,1518,772586,00.html.

of the solution to issues of constrained resources, rising demand and environmental concerns, let alone informed enough.

An easy way for regulators to address these discrepancies would be highlighting best practices among utilities. Falling short of demanding standards, regulators could showcase utilities offering their customers choices of renewable energy (where available), free energy efficiency consulting, high amounts of reimbursement for demand response, and information campaigns to keep rate payers informed and knowledgeable about the smart grid. Similarly to Japan's "Top Runner"[45] standard for appliances, which pegs efficiency standards to the most energy efficient product commercially available in a given appliance category, this "Smartest Utility" would set the benchmark for other utilities and allow customers to see what is available at a given time.

While informing rate payers about choices available is important, saving money is still one of the highest motivators for consumers considering changes to their energy usage behavior.[46] Yet as it stands, FERC has no power to regulate retail pricing for consumers and until now, there has been sporadic movement from state PUCs to introduce retail pricing. Adding financial incentives through measures such as real time pricing (RTP) will get a wider range of consumers involved and move some of smart grid's benefits within rate payers' reach. RTP will help smart grid deployment on many levels. It will allow customers to become active participants in the smart grid, paying attention to the dynamics of demand and supply and, with the help of smart appliances and measurement devices, adjusting their energy behavior accordingly. All eyes are currently on Ohio, where the state's Public Utilities Commission approved the nation's first trial residential real time price rate that automates consumer equipment response to RTP signals being tested under the ongoing AEP Ohio's gridSMART program.[47] Using transactive control, participating consumers' home

[45]See Komiyamaa, Ryoichi; Marnay, Chris. "Japan's Residential Energy Demand Outlook to 2030—Considering Energy Efficiency Standards 'Top-Runner Approach,'" Lawrence Berkeley National Laboratory, Berkeley, CA, May, 2008, available at http://eetd.lbl.gov/ea/emp/reports/lbnl-292e.pdf.

[46]Supra 43.

[47]Frequently Asked Questions, AEP Ohio gridSMART Demonstration Project, available at https://www.aepohio.com/save/SmartMeters/FAQ.aspx#4.

energy management systems will be linked with AEP's market-based dispatch system to communicate a cleared market price every five minutes.

Beyond engaging consumers, RTP is an essential enabler for vendors looking to help consumers make smarter energy choices through grid friendly appliances and home energy monitoring displays, among others. Without a price signal there will be no business case for these appliances and information management companies waiting to get involved.

On a larger level, RTP also is necessary to support the integration of renewable energy and make energy storage projects financially viable. Without a clear price signal that would allow cheap night-time wind energy to be stored and then sold at peak times during the day, these costly investments in renewables and large scale energy storage will not take place.

Smart Grid and the Workforce

Smart Grid has the potential to support the growth of a strong, highly skilled workforce in the United States. Recent private and public smart grid investments in the United States are estimated to directly create more than 280,000 jobs between 2009 and 2012, of which half are expected to be retained as permanent, high-skilled jobs.[48]

Yet without taking appropriate measures on the policy making level, there will be a severe disconnect between the specialization required and the skills available. First, there is the challenge of retaining the meter reader, or better yet, turning her into a meter technician. The experience of the electricity sector in countries in the EU-15, where 250,000 jobs have been lost since 1995, highlights the need for proactive measures.[49] Policymakers need to ensure that technological advances go hand in hand with educational excellence, leading to

[48]Supra 29.

[49]European Foundation for the Improvement of Living and Working Conditions (EUROFUND) (2008), *Trends and drivers of change in the European energy sector: Mapping report*, available at http://www.eurofound.europa.eu/pubdocs/2008/12/en/1/ef0812en.pdf, accessed August 2011.

the development of a smart grid-savvy labor force. Support for work force development, through vocational and on the job training and incentives for students choosing, for example, electrical engineering programs will be necessary to bridge the widening skills gap. Given the importance of involving consumers, incorporating a more customer centric approach into curricula for energy sector jobs also is advisable.

A second challenge that utilities across the nation face is a graying workforce. With an average age of 48 years, the US utility worker is five years older than the average worker. Worse still, between 25-35% of utilities' technical workforce is bound to retire during the next five years.[50] Given the prospects of thousands of engineers and technicians needed to manage the upcoming smart grid, these numbers are daunting.

There are examples of utilities and the government working together to mitigate these challenges. In 2010, the Obama Administration set aside about $2 billion for funding community colleges. Utilities, such as California's Pacific Gas & Electric (PG&E) are working with community colleges to develop curricula that support targeted workforce training across the Golden State. PG&E's PowerPathway program has helped 200 students to complete training in community colleges and subsequently hired half of those enrolled.[51]

As this example illustrates, the labor market impact of deploying smart grid needs to be a part of a new, holistic federal energy policy. Policymakers need to address these challenges by working with energy operators and educational institutions to realize smart grid's jobs potential. Without targeted collaboration, the intergenerational transfer of knowledge, and the transformation to a highly trained workforce, able to design, install, maintain and service smart grid infrastructure and technology will remain incomplete.

Further Steps

Further cooperation between the public and private sector is necessary to realize several of the smart grid's promises. Neither the gov-

---

[50]Supra 3, p.47.

[51]"Workers (and business) unite!," *The Economist*, Aug 27th 2011, p. 28.

ernment nor the energy operators alone can manage the expense associated with addressing the under investments in the U.S. power grid, where the average age of power transformers in service is more than 40 years.[52]

Policymakers can further budding public-private partnerships across the energy sector by examining areas where unbundling, restructuring and/or deregulation might unleash new markets that ultimately bolster the business case for smart grid.

As stated, smart grid in the United States is already blurring the previously clearly defined delineations between transmission and distribution. Reexamining the areas of state vs. federal regulation also could help utilities and grid operators with cost recovery while ensuring a wider deployment of smart grid. Regulatory innovation might, for example, help to level the playing in between states whose individual regulatory entities are currently using differing methods of review cost recovery for smart grid investments at the distribution level.[53] There is a need to examine if and how deregulation of the retail market can support smart grid deployment targets. And we need to explore new innovative regulatory approaches in unstructured markets that capture the flexibility of smart grid innovations in the current regulatory framework. While the Quadrennial Technology Review of the U.S. Department of Energy acknowledges that "deficient market structures" were one of the barriers to the deployment of energy storage technologies,[54] other studies of competitive markets have shown that unstructured markets "encourage the introduction of new products and services unavailable in more traditional electric markets."[55]

---

[52]Hill, Edwin; "New Challenges Demand New Solutions,"*EnergyBiz*September/October 2007, p. 14. http://energycentral.fileburst.com/EnergyBizOnline/2007-5-sep-oct/Financial_Front_New_Challenges.pdf.

[53]Supra 10, 38.

[54]Report on the First Quadrennial Technology Review, U.S. Department of Energy, September 2011, p.90. Available at http://energy.gov/downloads/report-first-quadrennial-technology-review.

[55] "Electricity "deregulation" inches forward, new report reveals," Smartgridnews.com, December 7, 2010, available athttp://www.smartgridnews.com/artman/publish/Business_Markets_Pricing/Electricity-deregulation-inches-forward-new-report-reveals-3349.html.

Why Cooperate?

While initial motivations about the smart grid's raison d'être are different in the United States and Europe, there are several overlaps in technological and regulatory challenges that call for closer cooperation. Both the United States and Europe are using the same technology and are tackling similar problems on the road to smart grid. Both could benefit from lessons learned so far, especially given the substantial investments in the United States following the 2009 stimulus funding. And, given the current economic situation in the United States and Europe, both can benefit from burden sharing in select areas of capacity building.

As the United States is examining its path towards smart grid, countries in the EU are confronted with several, similar policy choices that warrant a look at the benefits of transatlantic smart grid cooperation. On the topic of deregulation, countries in the EU and states within the US have experienced varying degrees of liberalized and deregulated electricity markets. In the United States, there are 14 states plus the District of Columbia with restructured electricity sectors.[56] States such as Texas are at the forefront of deregulating, allowing consumers to choose their electricity service from a variety of retail electric providers. The same is true for countries in the European Union, such as the United Kingdom. Comparing the experiences can help U.S. states with pending deregulation as well as European countries beginning the prescribed liberalization process to formulate and apply best practices. Deregulation is especially interesting to examine in light of ongoing smart grid technology deployments. Questions such as how deregulation affects smart grid deployment could be answered by looking at the experiences and challenges taking place on the other continent.

One challenge both the United States and Europe face is in the area of grid modeling. The smart grid of the future will employ millions of smart devices, smart loads and distributed renewable generation. The growth in smart sensors alone will create yet unseen volumes of data that need to be analyzed. At the same time, the forecasted expansion of intermittent renewable energy, especially from

---

[56]Supra 5.

small wind turbine and roof-top photovoltaic panels, will add a tremendous degree of uncertainty to planning and operating the grid as we know it.

While research[57] investigating future concepts and tools for grid operation and planning is underway, closer cooperation to tackle these challenges is needed. Researchers at Pacific Northwest National Laboratory have developed open-source simulation software called "GridLAB-D," which allows power systems planners and operators from the United States, Europe, and beyond to simulate the interplay of assets, control strategies, and communication devices over a time series and examine how a virtual smart grid would operate before money is spent on deployment.[58] These facilities allow for best practices to be shared while advancing the understanding of technologies and their impact.

There are several further areas of cooperation. Both the United States and Europe are staking great hopes in the promise of low-carbon electric vehicles. With plans to deploy millions of EVs across the continents, there are lessons to be learned in areas of grid friendly charging, electric vehicles as grid balancers through vehicle to grid technology, and much more.

Both the United States and Europe also are trying to integrate vast amounts of renewable energy coming online. Here, closer comparison between areas with substantial renewable energy in the electricity sector, such as the U.S. Pacific Northwest and EU countries Sweden, Denmark or Germany would be opportune to save grid operators from reinventing the wheel.

One area where cooperation already is successfully taking place is in the field of standards and interoperability. Under the helm of the Institute of Electrical and Electronics Engineers (IEEE), internationally harmonized data standards have benefitted vendors in both the United States and Europe to build applicable technology. Transferring IEEE's experience to smart grid specific applications, such as technology interfaces and communication needs could espouse similar positive outcomes. An organization that could spearhead this cooperation is the Smart Grid Interoperability Panel (SGIP), which coordinates

[57]See Future Power Grid Initiative, http://gridoptics.pnnl.gov/.

[58]http://www.gridlabd.org/.

standards development in the United States. Broadening SGIP's scope and including European standard organizations could create a forum to exchange and align ideas.

There already are several vehicles of EU-U.S. cooperation in place. The U.S.-EU Energy Council has two working groups on energy policy and energy technology RD&D that could be expanded upon to include lessons learned from EV deployment and renewable integration. As mentioned, the Smart Grid Interoperability Panel would be a prime location to include EU standards and interoperability discussions to facilitate smart grid deployment.

Comparing the policy driving smart grid in both the EU and the U.S. highlights the activist role the European Commission and Parliament are playing in defining smart grid and its deployment. While efforts by Brussels are often delayed and/or watered down, a key difference between the U.S. and the EU is that the EU actually has a smart grid deployment plan. Judging by the rapid and early deployment of smart meters and the incorporation of smart grid into larger political goals, such as emission reduction and electricity sector liberalization, the EU's smart grid outlook seems more concrete than in the United States at present. Taking a page out of the EU's playbook and developing a comprehensive approach to formulate smart grid deployment plans would benefit the grid modernization efforts in the United States.

The United States is well on its way to modernize its electric grid. Efforts driven by the legislative framework created by EISA and ARRA (and resulting funding) have tremendously helped to accelerate deployments of smart grid technology and infrastructure. Yet as recent evolutions in technology and societal demands for a more sustainable grid have changed the needs and characteristics of the power system, there is considerable room for regulatory and policy-making involvement to ensure the transition to a more efficient, resilient and smarter grid. This is especially important given the lessons learned from the progress made over the last four years.

While countries in the EU are undergoing a similar transformation, the U.S. experience in deployment and the resulting need to revisit the existing regulatory compact can provide vital insights to

European policy makers. At the same time, the United States, in its effort to formulate a forward looking federal energy policy, can learn from the progress that is being made in the EU, as well as the EU's efforts to establish guidelines to smart grid deployment across its 27 country-wide territory. Examining the others' experience especially in times of constrained resources will benefit both the United States and Europe. Mutual learning and further cooperation in smart grid deployment brings the promise of advancing the transatlantic relationship as both continents adapt to a new era of advanced technology, constrained resources and rising demand for energy.

Chapter Six

The Future of Renewable Energy

Reinis Aboltins

Transatlantic cooperation has considerable historical background: during the Cold War European and U.S. cooperation was based on shared security and military goals vis-a-vis a common adversary; common values; and similar approaches to democratic governance, political culture and ideology. With the change of global geopolitical relations and in a more multipolar world this traditionally close cooperation has relaxed in terms of attitudes, interests and goals in a growing number of situations. In this context transatlantic energy relations could actually become an example of ever-deepening cooperation. A number of factors point in this direction: energy has become by far one of the most important elements in international political and economic relations; volatility of energy resource prices makes energy-importing countries more vulnerable to various pressures from energy exporters; and global competition for energy resources has increased as economies in China, India and Brazil embark on an even faster pace of economic development. China has become the biggest energy consumer in the world, overtaking the United States. It has also become the biggest emitter of carbon dioxide. Thus it has also become the single most influential player in global energy relations that can affect global energy markets.

Global energy trends have affected both the U.S. and the EU, and the two players have also played key roles in setting energy trends globally. Only China has had equal or even greater impact on global processes. While the EU has taken the initiative among global actors in reducing greenhouse gas (GHG) emissions and becoming much more energy efficient, the U.S. leaves climate change and GHG issues more to the market, although it invests heavily in green technology development. The EU has set its ambitious 20-20-20 strategy as the cornerstone for the future development of its energy sector and economy in general.[1] Increasing the share of renewable energy has an essential role in this strategy.

There is a bigger framework for policies related to energy production and use—climate issues definitely play an important role in formulating policies, especially in long-term perspective. This is more obvious with the EU, but climate-related arguments also play a role in the U.S. energy debate.

Concerns about efficiency and the environment have been important overall drivers in global energy production and use. The motivations are not only ideological, but also economic—less efficiency and higher GHG emissions cost money. This trend characterises various levels of decision-making as well as a variety of situations in energy policy planning and energy industry, including production of primary energy resources as well as end use of energy. Natural gas power plants are replacing lignite and coal-fired power plants and smaller units are often replacing large centralized energy producing units in decentralized energy systems. For example, biomass combined heat and power plants (CHPs) of smaller capacity are replacing large centralized natural gas or coal CHPs.

Renewable energy accounted for 8% of total U.S. primary energy (TPE) consumption in 2010, the biggest proportion of which came from hydroelectric power, wood, biofuels and to a lesser extent from wind, waste, geothermal and solar/ PV.[2] For the EU the figure stood at 10.3% in 2008; the well-known target is 20% by 2020. The biggest contributors were hydropower, wind and wood.[3]

In terms of investment, the renewable energy sector is definitely a good market: the United Nations Environment Program (UNEP) reports that renewable energy investment worldwide rose to $211 billion in 2010.[4] While the majority of this funding goes to finance

---

[1] EU's "20-20-20" strategy is set to achieve a 20% cut in emissions of greenhouse gases by 2020, compared with 1990 levels; a 20% increase in the share of renewables in the energy mix; and a 20% increase in the energy efficiency.

[2] Hydroelectric power 2,5 quadrillion btu, wood 2,0 qbtu, biofuels 1,9 qbtu. Source: U.S. Energy Information Agency, 2011.

[3] Energy, transport and environment indicators, Eurostat, 2010, pp. 70-73.

[4] *Global Investments in Green Energy Up Nearly a Third to US$211 billion*, UNEP, July 7, 2011.

large-scale deployment projects rather than R&D or early commercialization activities, the level of financing indicates there is great interest in renewable and clean energy technologies as good financial investments. In 2010, venture investment in clean energy companies rose to $5.1 billion in the United States—23% of all venture capital investment for the year.[5]

Another shift is also notable: with investments of $72 billion in utility-scale renewable energy projects and companies, for the first time developing countries have spent more than developed economies on renewable energy. China has led this development; its nearly $50 billion invested in renewable energy in 2010 makes it by far the largest source of, and destination for, clean energy investment globally.[6] This clearly shows where future competition in renewable energy production will be coming from. Europe and the U.S. have to bear this in mind and act.

The Energy Security Context

It is inevitable that any debate on energy policy and development of energy infrastructure as well as on new technologies eventually involves discussions of energy security. The need for safe and economically viable supplies of energy resources and energy often sets the overall framework for any energy related debate. Volatility of oil and gas prices, political instability of suppliers of primary energy resources, and use of energy as a political tool make a grim set of factors affecting any energy importing country. The EU and the U.S. are big energy markets that produce energy on their own territories, yet each is significantly dependent on imports of primary energy resources, thus making security of supply a top priority. The EU has special relations with the Russian Federation, which is the biggest single natural gas supplier to the EU (approx. 40% of all natural gas imports) and the only natural gas supplier to a number of EU member states.[7] The U.S. is also among the biggest natural gas importers

[5] U.S. Department of Energy and EIA data, 2011.

[6] *Global Trends in Renewable Energy Investment 2011*, UNEP, Bloomberg New Energy Finance, 2011.

[7] The three Baltic states of Estonia, Latvia and Lithuania; and Finland. A number of other EU members are also significantly dependent on Russian gas supplies.

(combined both pipeline gas and LNG) in the world with 74 bcm in 2009, which is slightly less than Japan, Germany and Italy with 99, 83 and 75 bcm of natural gas respectively. The U.S. was the biggest importer of crude oil with 510 million tonnes in 2009, well ahead of China, Japan and India with 199, 179 and 159 Mt respectively[8].

The EU and the U.S. have established a fairly good framework for cooperation, including on energy issues. The EU–U.S. Energy Council (TEC) was created in November 2009 at the EU–U.S. summit, thus bringing the existing Strategic Energy Review to a higher level. The TEC's first meeting took place right after its establishment in November 2009, marking an important step in energy cooperation between the EU and the U.S. A common understanding was present at the meeting that energy security and climate change provide a broader context for energy policy on both sides of the Atlantic. Andris Piebalgs, EU Energy Commissioner at the time, pointed out that the Energy Council was a timely initiative in the context of growing global concerns on energy security and the important role that the energy sector has in climate change. The TEC's agenda covers such topics as energy security and energy markets, technology and R&D, and energy policy.

Other transatlantic forms of cooperation have also contributed to the EU–U.S. energy debate. The NATO Parliamentary Assembly noted in 2010 that energy and security are key variables in international security calculations.[9] Renewable energy sources can add comfortably to energy security in terms of a broader use of indigenous energy resources that many countries possess, but have not put to effective use so far. Energy security certainly requires consideration of a number of issues in a cross-disciplinary approach involving economic, social, technological, military, geological and environmental elements.

Another transatlantic forum for cooperation, the Transatlantic Business Dialogue (TABD), has provided a number of recommendations within the framework of the Transatlantic Innovation Dialogue, which

---

[8] *Key World Energy Statistics*, International Energy Agency, 2011, pp. 11-13.

[9] Jos Van Gennip, rapporteur, NATO Parliamentary Assembly, Committee Reports, 170 ESC 06 E—Energy Security.

is a spin-off of the Transatlantic Economic Council. The U.S. and the EU are the two largest economic zones in the world and innovation, R&D play an enormous role in fostering development of these two zones, which form probably the best market for innovation globally. This market provides a place for a multitude of players—researchers, companies and entrepreneurs—to collaborate effectively along a broad spectrum of topics including energy. And it is easy to see the huge potential for cooperation here—seeing the serious amounts of financing that the EU sets aside for innovation and technology only enhances what is already significant potential for cooperation. The EU's Strategic Energy Technology Plan (SET-Plan) has been set up by the European Commission to accelerate the development and deployment of cost-effective low carbon technologies. It comprises measures relating to planning, implementation, resources and international cooperation in the field of energy technology. The forecast budget for the SET-Plan amounts to €71.5 billion to be spent by 2050.

Energy efficiency and clean energy technologies are critical elements of the transatlantic economy and key aspects of strengthened transatlantic innovation on energy matters in ways that go beyond fossil energy supplies and physical security of energy infrastructure. TABD's recommendations point to a self-sustaining and climate-friendly global economy as a consequence of greater cooperation in innovation in the transatlantic market. TABD also notes that current export control regulations inhibit the flow of technology and technical information within the transatlantic economic area, subsequently creating barriers to coordinated transatlantic research and development.[10] Interestingly, while the U.S. Congress has not ratified such global climate documents as the Kyoto Protocol, TABD calls for the development and promotion of market-based common carbon accounting, which would facilitate more global trading of GHG emission credits than is currently done within the EU.

Regarding CO_2 emission credits, there are risks as well—for example, carbon prices might cast doubt on the sustainability of the carbon trading system in general. While several years ago it was predicted that carbon prices might rise as high as €60 per tonne of CO_2 emissions, the current price is around €10-15, which does not look very

[10] *Building a Transatlantic Innovation Economy*, February 3, 2010, p. 3.

motivating to move towards less emitting technologies for many CO_2 intensive industries and sectors of the economy. The EU emission trading system (ETS) obliges over 11,000 enterprises in the EU to buy emission allowances once they exceed quotas that are granted by the EU. Some industries are affected more than others. For example, airlines that are among the biggest emitters will have to find extra €1.12 billion to cover carbon costs in 2013, which will be the first year in which airlines must operate under the EU ETS system.[11]

The UNEP study on energy security risks shows that the strengthening of policy measures and the mitigation of energy security risks would benefit significantly from a strengthened and more coordinated multilateral producer-consumer dialogue between governments, industry, the financial community and relevant international organizations on a number of issues: (a) data and information sharing and increased transparency; (b) infrastructure investment and financing; (c) legal, regulatory and policy framework; (d) harmonization of standards and practices; (e) research, development and deployment of new technologies; and (f) investment/transit safeguards and burden sharing.[12] Most of these issues are very relevant when it comes to examining the potential for transatlantic cooperation on renewable energy.

A Framework for Thinking About Renewable Energy

There is general understanding in the EU and in the U.S. that more efficient use of resources and energy is a legitimate goal. There is also general understanding that adverse and inefficient use of energy resources does not facilitate sustainability of the energy industry and economic development, and can lead to depletion of the fossil energy resources that still form the backbone of the current global economic system. A shift towards a wider use of renewable energy resources is happening, even as fossil resources maintain a steady position through less polluting resources displacing more polluting fossil resources within total energy consumption.

[11]Airline Carbon Costs Take Off As EU Emissions Regulations Reach For The Skies, Standard & Poor's, February 18, 2011.

[12]George Kowalski and Sead Vilogorac, *Energy Security Risks and Risk Mitigation: An Overview*, UNECE Annual Report 2008, p. 83.

Geography and availability are of key importance when looking at renewable energy (RES) potential. The five renewable sources used most often are biomass, water (hydropower), geothermal, wind, and solar. Utilization of each of the five varies significantly depending on the region in question. In both the EU and the U.S. there are diverse examples of RES use according to regional principles. The potential for renewable energy resources also varies across industry, households and transport sectors. The transport sector, however, is the biggest energy consumer and thus offers considerable potential for RES use whether through biodiesel, bioethanol or electricity.[13] The transport sector accounted for 61.7% of oil use and only 1.6% of electricity consumption in 2009.[14]

Both the EU and the U.S. have measures in place that facilitate development of renewable energy. In the U.S., investment in and use of renewable energy is both encouraged and required by a range of federal, state, and local government legislation and utility incentives. The EU has ambitious goals of achieving 20% higher efficiency, 20% reduction of GHG emissions and 20% renewable energy by 2020 vis-à-vis 1990.

Signals for Energy and Energy Technology Markets

Long-term policy measures provide clear signals to a number of segments in the energy sector: technology production, primary energy production, as well as industrial and household consumption and everything related to this. All of these industries adjust their strategies and plans to such policy trends and decisions. If the policy emphasis is on improving energy efficiency, then technology producers are going to invest in improving the efficiency of technological processes of energy producing systems. If the policy emphasis is on developing wind power capacity, then producers can forecast the demand for new and more effective wind turbines and wind park management systems.

Going green demands investment, clear market stimulus and long-term policy vision. Even having all of the afore-mentioned does not

[13] *Key World Energy Statistics*, International Energy Agency, 2011, p. 33 and p. 37.

[14] Ibid., pp. 33-35.

guarantee an overnight switchover from fossil to renewable energy. There is no doubt that renewables will increase their market share of total energy consumption over the coming years, but they are not likely to displace the use of fossil fuels any time soon. This is because of much higher supply costs and requirement for vast tracts of land and water surfaces. Just to give an example: between 1990 and 2004, the contribution of renewables in meeting the total primary energy requirements of EU countries rose from 4.5% to 6.5%, from 12.0% to 14.5% for electricity generation, including hydro, and from 0.8% to 5.0% for electricity generation, excluding hydro. The corresponding numbers for North America show a decrease from 6.5% to 5.9% per cent for total primary energy, 18.6% to 15.3% for electricity generation, including hydro, and from 3.0% to 2.4% for electricity generation, excluding hydro.[15]

Infrastructure is one of the great challenges for the energy sector in terms of the scale of necessary action stemming from policies; the investment required to develop new infrastructure and renovate or replace the existing infrastructure (energy production, transmission and distribution/upstream and downstream); and the ever closer cooperation and synergies between a multitude of sectors of economy that are in one or another way related to energy, including energy consumption for industrial production.

Funding and Institutional Support

Both the EU and the U.S. have support measures in place for various energy types. Similarly both markets provide huge amounts of financing for research and development (U.S. Department of Energy (DoE) programs and EU Seventh Framework Programme (FP7)).

In the U.S. the Renewable Energy and Energy Efficiency (RE&EE) Advisory Committee was established by Secretary of Commerce Gary Locke in November 2010 as part of the Renewable Energy and Energy Efficiency Export Initiative, which has set a goal of doubling renewable energy exports in the next five years, targeting $10 billion or more in annual export benefits. In September 2011 RE&EE Advi-

---

[15] George Kowalski and Sead Vilogorac, *Energy Security Risks and Risk Mitigation: An Overview*, UNECE Annual Report 2008, p. 81.

sory Committee presented 11 recommendations to promote U.S. exports of renewable and efficiency technologies to federal officials.[16] The recommendations are the first set of actions the Committee was able to recommend to ensure robust growth in the renewable energy and energy efficiency exports by U.S. companies.[17]

In 2012 the total proposed budget for the U.S. DoE, which is the lead financial supporter of energy R&D in the United States, is $29.5 billion[18], with $3.2 billion going to the Office of Energy Efficiency and Renewable Energy (EERE) and $550 million to the Advanced Research Projects Agency–Energy (ARPA-E). This would represent an 11.8% increase over 2010. Critics point out, however, that compared to global investment in renewable energy research and development in 2010, when governments invested in renewable R&D alone more than $5 billion, the United States invested the same amount for all energy R&D during the same period.[19]

Additional energy R&D and early commercialization funding is also provided through tax benefits, grants, loans and contracts created by the American Recovery and Reinvestment Act (ARRA) of 2009. This stimulus legislation created $260 billion in energy tax credits for companies and consumers with the goal of improving the market penetration and share of efficient and clean energy technologies.

The EU's funding for research is channeled through its Framework Programme 7 (FP7), the total budget of which is approximately €50 billion from 2007 until 2013. Funding for energy research in 2011 was €216.9 million and will reach EUR €314 million in 2012 and an additional EUR €413 million during the last year of the current FP7. The total volume for energy research over the complete course of FP7 will amount to EUR €2.197 billion.

---

[16]See the full list of recommendations at http://export.gov/reee/eg_main_023040.asp.

[17]*Renewable Advisory Committee Delivers Recommendations to Under Secretary of Commerce and other Federal Officials*, Renewablenergyworld.com, September 14, 2011.

[18]See U.S. DoE "FY 2012 Statistical Table by Appropriation", http://www.cfo.doe.gov/budget/12budget/Content/Approstat.pdf.

[19]*Global Trends in Renewable Energy Investment 2011*, UNEP, Bloomberg New Energy Finance, 2011.

More than 30 states in the U.S. have financial incentives that subsidize the installation of renewable energy equipment. Net metering programs are in place in over 40 states allowing to pay for the net amount of electricity consumed, thus facilitating installation of micro-generating equipment for private use. Feed-in tariffs are widely used in the EU and also in a number of states in the U.S. specifically to support development and installation of certain types of renewable energy technologies.

Most of EU member states have legislation that sets framework and specific conditions for energy production from renewable sources along with energy efficiency measures. Priority issues in the EU are quite definitively related to the set 20-20-20 targets: energy efficiency, decrease of GHG emissions and a significant increase of renewable energy share in energy consumption. A very recent initiative of the European Commission foresees introduction of a new energy efficiency and savings directive. It has been put on the table because the Commission thinks that the current legislation does not provide the desired results and more stringent and compulsory measures have to be introduced to be able to achieve 20% higher energy efficiency by the target date. The new legislation, when adopted, would set binding energy efficiency measures for state and municipal institutions and large energy consumers. State and municipal institutions would be obliged to renovate 3% of their building stock annually and industrial consumers would have to undergo energy efficiency audits every three years. The Commission believes that these and other binding measures would help achieve the 20% energy efficiency and CO_2 emission reduction targets for 2020.

Climate and Economic Growth Dichotomy

Climate change and economic growth are two notions that are often interrelated in public rhetoric. Typically there are two schools with opposite views on the relationship between the two issues: one says that green energy imposes too much of a socio-economic burden (because green energy has to be subsidized in one or another way, but often through tariffs) and hampers economic growth (because green technologies cost more, demand higher investment and pay back over a longer time than fossil energy technologies); the other view propagates climate-friendly development of energy industry, the essence of which can be summarized under the green growth slogan. The EU pursues the rhetoric of climate change with generally higher enthusiasm than the U.S., where the "change" and "no change" dichotomy is more pronounced and controversial. However, regardless of ideologically motivated positions or the interests of industrial lobby groups on climate change, there is a general consensus that the least that can be done is being more effective in energy use.

There are energy industry sectors that do not exactly represent renewable energy as we understand it, but are very much in favor of achieving climate targets. The nuclear power industry is often mentioned as one potential solution to climate change processes. It is indeed true that nuclear power does not produce CO_2 emissions and thus can contribute to the reduction of CO_2 emissions. However, nuclear is not renewable, at least in traditional terms and it also has issues that do not have a solution currently—nuclear waste has to be buried somewhere and there is no guarantee that the depositories will remain intact for thousands of years to come. Finally, renewable energy is definitely a priority choice for the reduction of GHG and mitigation of the impact of energy production on climate.

Keywords: Competition and Cooperation

Global economic processes have contributed to both competition and cooperation between the EU and the U.S. Close economic and financial ties produce both opportunities and risks. This is true also about the EU–U.S. energy relations: competition and cooperation, opportunities and risks go hand in hand. On the one hand there has not been much discussion of European energy security in the United States, and thus the energy security debate centers around the possibility that the U.S. could be more self-sufficient in terms of energy supplies. On the other hand the recent eurozone crisis, with its risk of contagion to the United States, exposes the level of linkage between the two economies. For example, if a large-scale gas crisis were to occur in Europe partially shutting down production in Europe, then the U.S. also would be vulnerable to economic instability. Increasing transatlantic cooperation and ensuring European energy security would help to minimize the economic impact that such an adverse event would have on the U.S. Thus, in order to ensure its own economic security, the United States needs to bolster these efforts by strengthening energy cooperation with the European Union.

Both the EU and the U.S. invest heavily in energy-related research and development. The reason is quite clear: investment today in the development of better ways to handle extraction of energy resources, energy production or energy use is almost certain to lead to mid- and long-term gains. As commercially available fossil resources become scarce and renewables are growing in terms of sources and means of application, the efficient use of technologies used and organization of energy systems become not only more complex but more clearly geared to being sustainable. Some EU countries are keen to develop the ideology and principles of green growth, in essence emphasizing that economic growth does not have to come with depletion and or inefficient use of resources and or environmental pollution, e.g. greenhouse gas emissions, degradation of biodiversity, etc.

Energy security provides a dominating context for any further energy policy debate on both sides of the Atlantic. A good deal of discussion is centered around decreasing the dependence on imported fossil energy resources as well as more efficient use of energy resources. Nevertheless, it is clear that renewable energy will stay high on the policy agenda as an important means to increase energy independence from foreign supplies in a world of energy price volatility, particularly with regard to fossil resources.

The remaining question on the agenda is which renewables have the best effect on energy production in particular circumstances and how effective different kind of renewables can be. Natural gas is still better than biomass in terms of effectiveness and environmental impact.[20] In the U.S. the effectiveness of natural gas has facilitated use of shale gas in energy production, leaving questions only about the

[20]According to the U.S. Energy Information Agency CO_2 emissions in combustion for coal, natural gas and biomass are 960 kg/MWh, 596kg/MWh and 1509 kg/MWh respectively. "Carbon Dioxide Emissions from the Generation of Electric Power in the United States", US Department of Energy, Environmental Protection Agency, 2000, Table 1, http://www.eia.gov/cneaf/electricity/page/co2_report/co2emiss.pdf See also "Wood-Fueled Biomass Power Plants and CO_2 Emissions", Massachusetts Forest Watch, 2010, http://www.maforests.org/MFWCarb.pdf.

technology of extraction. On paper biomass has no CO_2 emissions, but in reality natural gas is 3-5 times more effective and produces 3 times less CO_2 emissions than biomass.[21] On the other hand, biomass is a resource that can be produced locally and is renewable: there is abundance of biomass available from a number of sectors like forestry and agriculture, to name but a few; and biomass CHPs can contribute to local economy through distributed power generation.

Expansion of the renewable segment goes hand in hand with improvement in production management systems and transmission of electricity. This is where modern power management comes into play in the form of deployment of smart grids, smart metering and good interconnections that are essential for managing an energy system with a multitude of various energy sources feeding energy into the grid. Distributed power generation and also microgeneration add to the complexity of tackling unstable wind power capacity and hydropower, which depend on the hydrological situation in their natural or artificial water reservoirs.

Further development of RES also depends on continued support, starting with R&D in the field of renewable energy and ending with managing grids with big renewable capacities connected. It is quite clear that deployment of smart grids in big and complex energy systems will take a lot of time and the existing power management and transmission systems need a lot of investment for being up-to-date anyway. With an increasing number of RES energy producing units connected to the grids, both the U.S. and EU member states are going to need massive investment in national transmission systems just to keep them up-to-date and able to handle varying power capacity. Technology producers in the EU and the U.S. have opportunities for competition and also for cooperation in this regard.

Renewable energy technologies similarly to some of the fossil technologies need relatively high upfront investment, therefore predictable and clear financial incentives have to be in place to make

[21]Figures may vary depending on technology and if life-cycle analysis of carbon intensity is applied. See, for example, "Findings from a time-dependent carbon footprint analysis of a biomass renewable energy plan", The Scottish Institute of Sustainable Technology, 2011, p.4, http://www.forthenergy.co.uk/pdf/Carbon%20Footprint%20Analysis.pdf.

renewable energy attractive to the investors. In addition to contributing to diversification of energy mix and tackling the climate issues, investments in RES have to produce payout. Furthermore, newer RES technologies have to become commercially available and viable. For example, estimates by the U.S. Department of Energy state that wind power has the potential to cover at least 20% of electricity production in the U.S. by year 2030, reaching 300GW.[22] Concentrated solar power, as well as photovoltaic electricity production, also has a very good potential to become an essential part of the national electricity production mix already by 2020 both in the U.S. and the European Union. Most renewable sources that are currently commercially available also contribute to reducing climate change stemming from energy production from fossil resources. At the same time, energy efficiency has to become a household phrase, because it makes little sense to produce energy from renewables if such energy is simply wasted.

Renewables are very much seen as part of the solution of energy security risks, along with clean coal and nuclear power. Oil and natural gas are certainly not to disappear from the energy menu, but might gradually give up part of their share. The U.S. and the EU are far from switching over from oil to renewable petrol in the transport sector, however, the share of biofuels is growing because of both market-driven choices and legislation requirements demanding an increase in biofuels in transport fuel mix. The role of electricity in transport is growing, also thus giving way to energy produced from renewable sources, be those wind, solar or hydro. Better commercial availability of microgeneration technologies combined with net metering and smart grids may well do the trick of more widespread use of electricity-powered vehicles.

Given the amount of funding for energy R&D and specifically renewable energy R&D on both sides of the Atlantic optimistic forecasts can be made about the potential for both competition and cooperation. While more competition can be expected on the commercial side, the regulatory and legislative framework set by decision makers

---

[22]Franklin Kramer and John Lyman, *Transatlantic Cooperation for Sustainable Energy Security*, The Atlantic Council and Center for Strategic and International Studies, February 2009, p. 31.

can contribute to ever closer cooperation between the two key players on the global energy stage.

Recommendations for Policymakers

Given both the competitive and cooperative potential of the Transatlantic relations and the increasing global competition for energy a number of conclusions and recommendations for policymakers in Europe and the United States can be drawn.

Global consumption of fossil fuels is growing and prices tend to get higher opening up opportunities for better use of domestic energy sources an important part of which is renewable energy. Clear policy goals and set targets give signals to the energy and related industries about the potential for investment and revenue down a foreseeable time-line.

Growing energy consumption in some of the biggest energy markets means an ever increasing competition for all kinds of energy that can be imported. In this light ability to be more energy efficient is essential opening up opportunity for development of energy efficient technologies both on the production and consumption side. Set targets in this respect allow economically based calculations on savings against investment.

While competition does not have to disappear there is most certainly a place for cooperation in research and development on renewable energy. Efficiency of renewable technologies is increasing rapidly thus facilitating proliferation of renewables in energy production. Creating more favorable conditions (legislation and infrastructure) for the use of renewables in micro-generation would definitely add to the overall capacity of non-industrial energy production.

Convergence of rules is probably not possible and not necessary, however, the positive potential of harmonization of certain export regulations related to technology export or exchange from both the U.S. as well as the EU should be welcome.

Intensified political dialogue on energy and energy technology cooperation will send a clear message to the market about the economic potential of cooperation and will add to the stability of forecasts on the future development of transatlantic energy and energy technology market that will also increase ability of the two players to compete with other regions and players worldwide.

Chapter Seven

Trading the Sun

Michael Stanton-Geddes

While solar power has enormous potential, it faces serious challenges. Heliophiles never cease to remind us that "within six hours deserts receive more energy from the sun than humankind consumes within a year,"[1] and throughout human history, engineers and researchers have attempted to utilize the sun's energy by turning solar irradiation into a medium that provides heating, cooling, and lighting services. However, today direct solar energy accounts for only about 0.1% of total primary energy supply world-wide, and the international community still speaks of solar energy as the "new energy" to power the world tomorrow. Is it time to give up?

This chapter argues that the obdurate challenges are socio-political, not technological. In the past decade, solar energy returned to the front pages, particularly with growing concerns about climate change, sustainable development, as well as the most recent oil price spike in 2006. At the same time, the solar industry is on the verge of achieving its long-predicted potential. Alas, history has shown that interest in solar energy arrives in waves, often linked to energy price shocks, and that these waves crest and their momentum dissipates. The EU and U.S. should make use of the momentum as well as newest technological advances and adopt a new joint approach to solar energy research while simultaneously opening global markets to solar energy. In the midst of a prolonged, global economic slump, as public interest in climate change wanes,[2] this task will be difficult, but the economic and environmental returns warrant the investment.

---

[1] Dr. Gerhard Kies, DESERTEC Foundation, quoted on DESERTEC website homepage. Accessed September 8, 2011 at http://www.desertec.org/.

[2] "Global Concern for Climate Change Dips Amid Other Environmental and Economic Concerns," Nielsen Company (New York: August 29, 2011). Available at

The environment, energy security and economy—these are the three key considerations that drive public support for renewable energy technologies. For many years, politicians and advocates of green energy included the economic benefits of solar technology as an additional factor. Introducing the 2007 bill in the United States Senate on climate change, U.S. Senators Lieberman and Warner "presented their new bill as the core of a new federal program that Congress should pass to avert catastrophic global climate change while enhancing America's energy security," and only once mentioned jobs.[3] Since 2008, worsening economic conditions globally have strengthened and juiced the economic argument. In 2010, Senator Lieberman released a new bill, stating that the "comprehensive energy and climate change legislation will create jobs, strengthen America's energy independence, safeguard our national security, and restore our global economic leadership for decades to come."[4] Europe underwent the same transformation, from its Climate and Energy package of 2008 to the 2010 State of the Union address, in which European Commission President José Manuel Barroso said the EU must continue "to deliver on climate and energy package, as a core driver for change [...] this will not only strengthen our economy tomorrow: it will provide new openings today."[5] Politicians today have changed the order of importance, and rank the economic objectives of renewable energy development first.

The international climate and domestic economy dimensions can contradict each other. On one hand, the rapid and broad deployment

---

http://blog.nielsen.com/nielsenwire/consumer/global-concern-for-climate-change-dips-amid-other-environmental-and-economic-concerns/.

[3] "Lieberman and Warner Introduce Bipartisan Climate Legislation," Press Release, Office of Senator Joe Lieberman (Washington, DC: October 18, 2007). Available at http://lieberman.senate.gov/index.cfm/news-events/news/2007/10/lieberman-and-warner-introduce-bipartisan-climate-legislation.

[4] "Kerry, Lieberman: American Power Act Bill Will Secure America's Energy, Climate Future," Press Release, Office of Senator Joe Lieberman (Washington, DC: May 12, 2010). Available at http://lieberman.senate.gov/index.cfm/news-events/news/2010/5/kerry-lieberman-american-power-act-bill-will-secure-americas-energy-climate-future.

[5] José Manuel Durão Barroso (President of the European Commission), *State of the Union 2010*, European Commission (Strasbourg: September 7, 2010). Available at http://europa.eu/rapid/pressReleasesAction.do?reference=SPEECH/10/411.

of low-carbon technologies requires low barriers and liberalized trade. On the other hand, maintaining clean-energy technology manufacturing creates temptations for protection. Governments face strong political pressure to ensure that the budget expended on subsidies and support for renewable energy goes to domestic producers. If there is to be conflict across the Atlantic Ocean, it will be on the protection of these industries against international trade. The first salvo hit the transatlantic relationship in August 2011. The European Commission requested consultations with the government of Canada under the auspices of the WTO regarding the feed-in-tariff (FIT) program in Ottawa, which included local content requirements of 50% for wind projects and 60% for solar projects by 2012, arguing that this weakened the ability of European exporters to sell equipment in Ottawa.[6] These squabbles signal a perilous path into trade conflicts.

This chapter argues in favor of a coordinated approach to solar energy research and trade. Section one describes the status of the solar energy technology. Section two describes common policy approaches in the United States and Europe regarding solar power and renewable energy. Section three examines the rise of the 'green growth' and 'green jobs' objectives to the top of the agenda and section four reveals contentious points between the U.S., EU and global partners, as they balance environmental objectives and economic objectives. These sections support the argument for increased collaboration across the Atlantic for solar energy and solar power development.

Solar Energy Technological Status

In the modern era, it is too short-sighted to look back to climate change awareness resulting in the 1997 Kyoto Protocol, or President Carter's installation of solar panels on the White House roof in 1978 as the beginning of the solar energy era. Public support of solar technology moves in waves of interest and apathy.

---

[6] "The EU requests WTO consultations with Canada over Ontario's renewable energy policy," *European Commission, Trade Directorate-General* (Brussels, Belgium: August 11, 2011). Available at http://trade.ec.europa.eu/doclib/press/index.cfm?id=732 . WTO Dispute record for Dispute DS412, "Canada—Certain Measures Affecting the Renewable Energy Generation Sector, available here: http://www.wto.org/english/tratop_e/dispu_e/cases_e/ds412_e.htm.

Following World War II, the first modern age of global interest in solar energy involved a host of new international organizations. The United Nations accredited the International Solar Energy Society (ISES) in 1954 to support the advancement of solar energy, and in 1958 the UN's Department of Economic and Social Affairs published a study about solar power, *New Sources of Energy and Economic Development*. From the 1950s onwards, UNESCO organized conferences about solar power in the United States, India and France. In the mid-1950s, at a Solar Energy Laboratory in the French Pyrenees, a parabolic mirror was used to generate 30 kW of power.[7] Despite this activity, interest faded; oil cost little and nuclear grew fashionable.

The 1973 and 1978 oil shocks, as well as growing discomfort with nuclear power, changed the situation. Tam Dalyell, a Member of the European Parliament and the rapporteur of its energy budget, wrote in *New Scientist* in October 1978 that "the most serious work is being done in Europe, probably in the world, on solar energy" at the European Community's research facility in Ispra, Italy.[8] In 1977, the U.S. government established a short-lived Solar Energy Research Institute (SERI), and respected research groups such as the Worldwatch Institute predicted that, by the year 2000, solar energy would provide a quarter of energy.[9] Nevertheless, for a second time, oil prices decreased and postponed this event.

In the past decade, solar energy returned to the front pages. The UN's Department of Economic and Social Affairs states that "New and renewable sources of energy have received a great deal of attention since the World Summit on Sustainable Development was held in Johannesburg in 2002."[10] Others would argue that attention returned

[7] *UNESCO Courier*, UNESCO, XI, 9 (Paris: 1958). Daniel Berhman, "The Sun at Work around the World," *The UNESCO Courier: a window open on the world*; XI, 9 (Paris: 1958), pp. 1-17. Available http://www.unesco.org/new/en/unesco-courier/archives/.

[8] Tom Dalyell, "Westminster Scene," *New Scientist*, Vol. 80, No. 1125 (October 19, 1978), p. 211.

[9] Denis Hayes, *Worldwatch Paper #11: Energy: The Solar Prospect*, Worldwatch Institute (Washington: 1977).

[10] United Nations Department of Economic and Social Affairs, Renewable Energy website. Accessed October 21, 2011 at http://www.un.org/en/development/desa/climate-change/renewable-energy.shtml.

in earnest with concern about climate change in the mid-1990s, or the oil price spike of 2006. Solar energy technology rides a supportive wave, but the political controversy over the bankruptcy in August 2011 of a U.S. firm, Solyndra, which received a loan from the U.S. Department of Energy, hints that the momentum of the wave is dissipating.

This chapter largely focuses on efforts to generate electricity from solar power because energy analysts and national and international energy statistics bureaus anticipate the sector to play a role in meeting future electricity demand. Today, coal, gas and oil fired power plants provide 67% of electricity worldwide (nearly 70% in the United States, and 48% in the eurozone).[11] Consumers in advanced economies use more electricity, and rural and universal electrification projects increase access to people in developing regions, so the demand for new power plants and electricity generation capacity is increasing.

Unlike other types of thermal power generation, solar power applications, especially PV, can be used in many configurations. Some are small enough to provide energy to one appliance, such as a handheld calculate, or just one household. Off-grid systems operate well in remote areas to provide electricity for specified, limited services, such as medical clinics or schools. Large centralized solar PV and CSP power stations function as full scale utilities, providing many megawatts of power.

Three technological pathways exist to convert the sun's rays into energy: photovoltaic (PV) cells, which directly convert solar energy into electricity; concentrating solar power (CSP), which uses mirrors to reflects the sun's rays and heat medium, making steam to turn turbines and generate electricity; and solar fuels, in the early stages of basic research, which apply solar energy to chemicals or other compounds to produce fuel.

[11]Electricity production from oil, gas and coal sources (% of total), World Bank Data (series EG.ELC.FOSL.ZS). Available at http://data.worldbank.org/topic/energy-and-mining.

Solar Energy Technologies

Concentrating Solar Power (CSP)

Concentrating Solar Power utilizes an array of mirrors to reflect and concentrate the sun's rays at one point, rapidly creating temperatures in the thousands of degrees. A heat transfer medium, such as water, molten salt, or oil, produces steam to turn a traditional turbine and generate electricity. Scientists and researchers have developed CSP technologies for well over a century. Today, four categories account for all CSP systems: parabolic trough, liner Fresnel mirrors, central tower, and dish systems. California installed the first modern commercial CSP plant for electricity generation from 1985 to 1991. Concentrating solar power electricity generation is best suited in areas with clear skies and direct sun—the "Sun Belt" across southwestern United States, southern Europe, North Africa and the Middle East, and Australia.

Photovoltaic (PV)

Photovoltaic cells exploit the properties of semiconducting materials, such as silicon, to generate a voltage as solar irradiance crosses the semiconductor. A wide-range of photovoltaic technologies exists, with silicon cells dominating the market. The U.S. government utilized solar cells in satellites and defense applications in the 1950s. RCA Laboratories introduced the first commercially successfully photovoltaic cell, the a-Si solar cell, in 1976. Today, emerging PV technologies at the basic research phase include dye-sensitized solar cells, organic PV cells, advanced semiconductors and spectrum converters. The best PV cells achieve efficiencies around 25 percent.

Solar Fuels

The third and most immature solar energy technology is solar fuel. Using either electricity generated from PV/CSP or direct solar heat and a chemical or biological process, scientists have been able to produce fuels, such as methanol, diesel, hydrogen and mixed gases. Most research is still in the basic phase, so this paper focuses on commercial solar technologies.

Source: Many papers discuss the types of solar energy in far more detail. The best, most comprehensive source is the IPCC's Special Report on Renewable Energy Sources and Climate Change Mitigation (SRREN).[1]

[1]Arvizu, D., P. Balaya, L. Cabeza, T. Hollands, A. Jäger-Waldau, M. Kondo, C. Konseibo, V. Meleshko, W. Stein, Y. Tamaura, H. Xu, R. Zilles, Direct Solar Energy in O. Edenhofer, R. Pichs-Madruga, Y. Sokona, K. Seyboth, P. Matschoss, S. Kadner, T. Zwickel, P. Eickemeier, G. Hansen, S. Schlömer, C. v. Stechow (eds), IPCC Special Report on Renewable Energy Sources and Climate Change Mitigation, (Cambridge University Press, Cambridge, UK, 2011).

Policies to Promote Solar Power

A well-developed literature on innovation economics demonstrates the importance of government support for invention and innovation.[12] Governments and private investors have many options to support new technologies along the simple and effective model of the innovation process below. At the front-end, or initial stages of basic research, economists have demonstrated that markets invest below the equilibrium level for optimal social gain because of factors including risk, uncertainty, increasing returns to scale and failure to capture all gains. As a result, governments should fund laboratories and academic institutions to fill the investment gap.

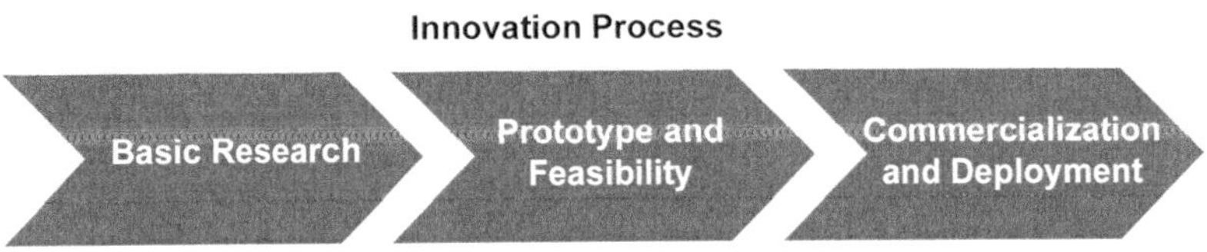

At the back-end of the innovation sequence—commercialization and deployment—governments play an important role in market regulation and creating or subsidizing markets. Amongst the many back-end policies that serve to promote renewable energy, two policy families have driven the deployment of solar electricity generation. Renewable portfolio standards (RPS) mandate that renewable energy sources comprise a certain share of the total energy mix. A renewable portfolio standard is technology-neutral in theory, though rarely in implementation. Some RPS policies specify that a certain share must come from solar power. The second category, feed-in tariff (FIT) policies, provide a fixed, long-term and guaranteed tariff or a subsidy per kWh to a producer of solar

[12]See Richard R. Nelson, *National Systems of Innovation* (Oxford, England: Oxford University Press, 1993); Lewis Branscomb and Phillip Auerswald, *Between Invention and Innovation, An Analysis of Funding for Early-State Technology Development* NIST GCR 02-841 (Washington: National Institute of Standards and Technology, 2002); George S. Ford, Thomas M. Koutsky and Lawrence J. Spiway, *A Valley Of Death In The Innovation Sequence: An Economic Investigation*, Discussion Paper prepared for the Commerce Department, Technology Administration (Washington, DC: Phoenix Center for Legal and Economic Public Policy Studies, 2007).

electricity. Feed-in tariffs necessarily require a qualification or certifications of facilities to receive the tariff.[13]

Europe

The European Union (EU) provides front-end support to assist researchers in solving technological challenges, as well as back-end incentives to level the playing field between solar electricity and carbon-based generation. EU member states have pursued independent energy policies for many years, working on solar and other renewable technologies, and national agencies retain strong control over research priorities and budgets. Italy, Spain and Germany accounted for 95% of member state spending, and companies in those countries also invested the most in R&D, reflecting a correlation between research activity and government support. Spain's Center for Energy, Environment and Technological Research operates a solar power prototype site at Almeria, which is the largest European center for research, development and testing of concentrating solar power. Germany, another example, directs research through its Federal Program on Energy Research and Energy Technology's New Energy area. Multiple ministries, including the Ministry for Environment and of Education and Research, provide funding to activities all along the innovation pathway, from basic research to deployment of prototypes, advancing German industry's technological leadership.[14]

While member states continue to maintain autonomy over their research priorities, the European Commission, especially since the creation of a Directorate for Innovation in 2009, aims to streamline and enhance energy research. The EU's research flagship is the Seventh Framework Programme for Research and Technological Development (FP7), which runs from 2007 to 2013 in this iteration. FP7 is funding 29 PV research projects for a total of €106 and six CSP proj-

[13]Toby D. Couture, Karlynn Cory, Claire Kreycik and Emily William. *A Policymaker's Guide to Feed-in Tariff Policy Design*, Technical Paper NREL/TP-6A2-44849 (Golden, Colorado: National Renewable Energy Laboratory, US Department of Energy, 2010).

[14]*Photovoltaic Power Systems Programme (PVPS) Trends Report 2009*, (Paris: International Energy Agency, 2009), p. 21. Available at http://www.iea-pvps.org/index.php?id=32.

ects for a total of €21 million, as of October 2011.[15] In the previous Framework Programme from 2002 to 2006, the EU dedicated €110 million to 30 solar projects, covering a range of technologies and processes. Member states and the private sector funded another €350 million of research into photovoltaics, with corporate R&D accounting for nearly 60% of the total.[16] The EU dedicated much less funding to CSP research, a total of €86 million in 2007. The EU provided about one-quarter of the funding in 2007 (€20 million), member states a little less, and corporations funded 56% of total R&D in CSP.[17]

The European Commission's Directorate-General for Energy in 2009 created the Strategic Energy Technology Plan (SET-Plan), to reduce redundant activity, combine expertise, and share costs across many member states. Under the SET-Plan, the European Commission created the *Solar Europe Industrial Initiative* in 2008, an opportunity for European companies to increase their innovation base and improve their competitiveness. The Solar Europe Industrial Initiative exists to "focus and align the efforts of the Community, member states and industry in order to achieve common goals and to create a critical mass of activities and actors, thereby strengthening industrial energy research and innovation on technologies for which working at the Community level will add most value."[18] Specifically, the SEII designed the technology roadmap and guides progress towards milestones for demonstration and deployment of both solar photovoltaic (PV)[19] and concentrating solar power (CSP).[20]

[15] "EU Energy Research Projects," Directorate General for Research & Innovation, database (Brussels: European Commission, 2011). http://ec.europa.eu/research/energy/eu/projects/index_en.cfm#results, accessed on October 20, 2011.

[16] "Photovoltaics," Strategic Energy Technologies Information System (Brussels: European Commission, 2011). http://setis.ec. europa.eu/newsroom-items-folder/capacities-map-photovoltaics, accessed on October 20, 2011.

[17] *Concentrated Solar Power*, Strategic Energy Technologies Information System (Brussels: European Commission, 2011). http://setis.ec. europa.eu/newsroom-items-folder/concentrated-solar-power, accessed on October 20, 2011.

[18] *Industrial Initiatives*, Strategic Energy Technologies Information System (Brussels: European Commission, 2011). http://setis.ec.europa.eu/activities/initiatives, accessed on October 20, 2011.

[19] *Solar Europe Industry Initiative Implementation Plan 2010-2012*, Solar Europe Initiative (Brussels: European Photovoltaic Industry Association (EPIA) and PhotoVoltaic Technology Platform, May 2010). Available at http://ec.europa.eu/energy/technol-

Up to 2020, the EU plans to invest €16 billion in solar energy research, leveraging public funds to attract private financing in public-private partnerships that share risks, and rewards, in the development of solar energy technologies.[21] Surprisingly, the EU did not fund a single solar energy project in the €3.8 billion European Economic Recovery Programme, though it is not clear if this is because solar was seen as unproductive for investment, or already sufficiently supported through other programs.[22]

On the back-end, the EU leads the world in meaningful policy action. The EU and its member states are using market-pull policies to support the development of solar energy, notably generous FIT policies, a RPS and a greenhouse gas reduction mandate. The EU adopted in 2009 a comprehensive energy and environment plan (20-20-20), which includes an effort-sharing decision for the EU collectively to reduce carbon dioxide (CO_2) 20% by 2020 compared to 1990 levels, a binding target for collective use of 20% renewable energy, also by 2020, and a strengthened and expanded emissions trading scheme (ETS) emissions to account for the climate cost. Though the EU's RPS is largely technology-neutral, the European Commission forecasts envision solar power providing up to 20 percent of the portfolio standards.

Spain, Germany, France, Italy and the Czech Republic implemented FIT policies. For example, Spain's 1997 electricity law created a feed-in tariff of 80-90% of the market price, Percentage-based FIT models did not always support solar because until recently, solar

---

ogy/initiatives/doc/pv_implementation_plan_final.pdf.

[20] *Solar Thermal Electricity European Industrial Initiative (STE-EII) Implementing Plan 2010-2012*, Solar Europe Initiative (Brussels: European Solar Thermal Electricity Association, May 2010). Available at http://ec.europa.eu/energy/technology/initiatives/doc/csp_implementation_plan_final.pdf .

[21] "Joint Statement on the launch of the European Wind, Solar, Electricity Grids and Carbon Capture and Storage Industrial Initiatives" (Brussels: European Commission, Presidency of the European Union, SET-Plan Steering Group, June 3, 2010). Available at http://ec.europa.eu/research/energy/eu/policy/set-plan/joint_statement_for_the_eiis_launch.pdf .

[22] "EU Energy Projects Funded Under The European Economic Recovery Programme," European Commission. Available at http://ec.europa.eu/europe2020/pdf/energy_project_en.pdf.

energy cost 100% more than market prices, so even with a 100% feed-in tariff, the solar power producer could not reach cost recovery. Consequently, Spain revised its FIT 2006 to create a fixed price above retail, or a sliding premium.[23] Responding to rapid subscription and large costs, estimated at $26.4 billion in solar energy FIT payments for 2008 alone, the Spanish government capped the program, and the market collapsed. New legislation in 2010 removed premium options for CSP, and adds a cap to the number of hours per year that existing solar projects receive FIT payments, weakening these projects financial position. The boom and bust of Spain's solar energy industry due to the FIT policy provides important lessons for designers of FIT policies in the future.

The United States

The United States has deliberately supported science and technology at the federal level since the end of World War II, guided by Vannevar Bush's call to President Truman for the United States to maintain the vigorous "pioneer spirit" in the realm of science.[24] For solar power, the United States built a complex and comprehensive system of front-end, or technology-push, but has failed to enact market measures to develop demand for solar power.

The U.S. government provides the majority of its support for solar power through front-end policies, providing sixteen financial incentives to solar power, at least three regulatory incentives, and research funding through the Departments of Energy, Defense, Agriculture and Commerce. The DOE's Solar Program provided $247 million for solar energy technology in 2010.[25] The National Renewable Energy Laboratory, the doyenne of the Department of Energy's national laboratories, conducts and directs substantial research in solar energy. A host of other institutions conduct basic research, partner with universities and private companies, or provide loan guarantees and federal

[23]Couture, et. al., *A Policymaker's Guide to Feed-in Tariff Policy Design*, p. 20.

[24]Vannevar Bush, *Science: The Endless Frontier* (Washington, DC: Government Printing Office, 1945). Available at http://www.nsf.gov/od/lpa/nsf50/vbush1945.htm.

[25] "Solar Energy Technologies Program," U.S. Department of Energy, Energy Efficiency & Renewable Energy (EERE). http://www1. eere.energy.gov/office_eere/pdfs/51841.pdf, accessed on October 20, 2011.

investments in solar power research.[26] The most recent is the Advanced Research Projects Agency—Energy (ARPA-E), which provides funding for high-risk, high-reward projects, with the objective of filling the under-investment gap discussed above. In February 2011, the Department restructured solar research under the "SunShot Initiative" which aims to "drive widespread, large-scale adoption of this renewable energy technology and restore U.S. leadership in the global clean energy race." SunShot is analogous to the EU's Solar Industrial Initiative, focusing on the commercialization of solar energy technologies. It primarily aims to bridge the 'valley of death' in the innovation process between basic research and deployment.

The only back-end, market-based measure in the United States at the federal level to support the deployment of renewable energy and solar power is a performance-based incentive, the Renewable Energy Production Incentive (REPI), in force since 1992, which provides a maximum of 2.2¢/kWh to qualifying renewable energy producers, though it frequently pays less because of fiscal constraints. The United States enacted the first major feed-in tariff policy in the world, in the 1978 Public Utility Regulatory Policies Act (PURPA), which encouraged renewable energy in response to the oil price shock, and required utilities to purchase power from other independent power providers. Power sector deregulation and lower primary energy source prices diminished the impact of PURPA. U.S. legislators proposed bills over the past five years to enact a federal RPS or carbon market, notably the Lieberman-Warner 2007 (America's Climate Security Act) and the Lieberman-Kerry 2010 (American Power Act), yet the government has not passed any form of national legislation to mandate greenhouse gas emission decreases, require renewable energy use, or place a price on carbon.

In the absence of federal action, U.S. states, led from the West by California, have enacted back-end policies of their own to create the marginal markets for solar energy, primarily Feed-In Tariffs and Renewable Portfolio Standards. As of September 2011, California, Hawaii, Maine, Oregon, Rhode Island and Vermont, in addition to

---

[26] "Fact Sheet: Department of Energy Investments in Solar Energy," U.S. Department of Energy, Energy Efficiency & Renewable Energy (EERE). http://www.eere.energy.gov/pdfs/fact_sheet_doe_investments_in_solar.pdf, accessed on October 20, 2011.

municipalities in California, Florida, Indiana and Texas have enacted FIT legislation.[27] 38 states, Guam, the Northern Mariana Islands and the District of Columbia enacted mandatory renewables portfolio standards, joined by four municipalities or cities.[28] Unlike in Europe, where member state level legislation applies EU law, in the U.S state laws often lead federal law.

Industry Development

Relative to other sources of energy, solar power is practically invisible. Globally, all forms of solar energy accounted for 0.1% of total primary energy supply in 2008. Even among renewable energy sources, which provide 12.9% of the world's primary energy supply, solar power accounts for about 0.8% (biomass provides the lion's share, over 80%).[29] Solar energy remains extremely limited compared to other energy sources.

Solar power is growing as PV and CSP costs decrease and government policies support the sectors' commercialization, however, solar power is still more expensive than any other form of electricity generation; the levelized cost of electricity calculated by the IEA still exceeds the high-end cost estimates for coal, gas and nuclear, even when including a carbon price of $30/ton, a price which exceeds the average market price in Europe.

---

[27] "Database of State Incentives for Renewables & Efficiency, Solar (DSIRESOLAR)," N.C. Solar Center, N.C. State University and the Interstate Renewable Energy Council. http://www.dsireusa.org/incentives/ index. cfm?EE=1&RE=1&SPV=0&ST=0&searchtype=RPS&technology=all_solar&sh=1, accessed on October 21, 2011.

[28]DSIRE Solar.

[29]*Summary for Policy Makers* in: O. Edenhofer, R. Pichs_Madruga, Y. Sokona, K. Seyboth, P. Matschoss, S. Kadner, T. Zwickel, P. Eickemeier, G. Hansen, S. Schlömer, C. von Stechow (eds), *IPCC Special Report on Renewable Energy Sources and Climate Change Mitigation* (Cambridge University Press, Cambridge, United Kingdom and New York, NY, USA, 2011), p. 6. Available at http://srren.ipcc-wg3.de/report/IPCC_SRREN_SPM.

Table 1. Levelised Cost of Electricity (US$ per MWh)

Technology	Low-end	High-end	Investment costs (%)	Fuel costs (%)	O & M (%)	Carbon costs (%)
Coal	67	142	42	23	8	27
Gas	76	120	16	67	5	11
Nuclear	42	137	75	9	15	0
Wind (onshore)	70	234	87	0	13	0
Wind (offshore)	146	261	80	0	20	0
Solar (CSP)	136	243	n/a	0	n/a	0
Solar (PV)	215	600	n/a	0	n/a	0

Assumptions: 10% discount rate for all technologies. Coal prices of 20 USD/ton, gas price of USD 10.3/MMBtu in OECD Europe and USD 11.7/MMBtu in OECD Asia. Carbon price of 30 USD/ton.
Source: *Projected Costs of Generating Electricity* (2010). International Energy Agency (Organisation for Economic Co-operation and Development/International Energy Agency, 9 rue de la Fédération, 75739 Paris Cedex 15, France).

PV Industry

PV power production reached 40 GW at the end of 2010, after installers added 16.7 GW, a 73 percent increase over 2009. Four countries dominate the PV solar power sector: Germany, which alone accounts for 43% of installed global capacity; Spain, accounting for another 10%; Japan and Italy, both around 9%; and the United States at 6%. Spain, Italy, France and the Czech Republic recently joined the exclusive group of countries with more than 1 GW of installed solar capacity, all increasing their installed capacity many thousand-fold between 2005 and 2010.[30]

Germany is the leader in absolute terms of installed PV capacity, but other countries rapidly increased their capacity. However, growth rates for this sector are misleading. For example, Italy in 2010 had only 20% of Germany's total installed capacity, Italy's astronomical growth rate of 9239% from 2005 to 2010 increased capacity from only 37.5 MW to 3502 MW, whereas Germany's more sedate 742% increase over the same period added four times more cumulative installed capacity (15.3 GW compared to 3.5 GW for Italy). The chart

30 "Renewable Energy—Solar (Installed capacity)," *BP Statistical Review of World Energy 2011*. Available at http://www.bp.com/sectionbodycopy.do?categoryId= 7500&contentId=7068481.

Figure 2. PV Solar Power—Cumulative Installed Capacity (MW) and Increase 2005–2010

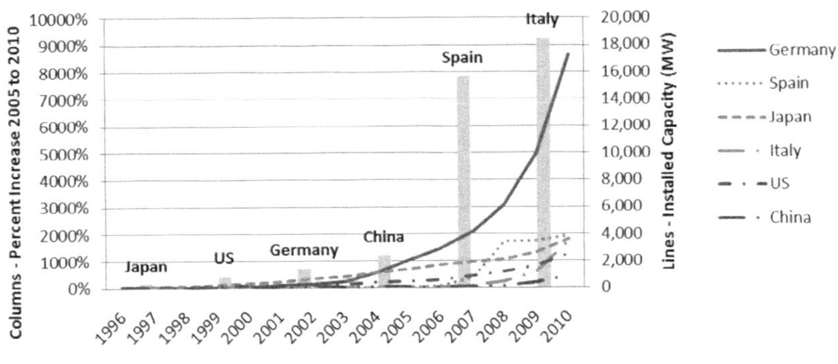

Source: *BP Statistical Review of World Energy*, June 2011.

above demonstrates the sky-high growth rates and also the absolute numbers to provide perspective.

The United States increased total installed PV capacity by 426% since 2005, and Asia's growth of installed solar capacity averaged 633% from 2005, with South Korea increasing PV capacity by 4144% and China by 1213%.[31]

The PV sector is developing into a global industry, with competition for resources and markets, especially the PV sector. The largest Solar PV manufacturers by market share are First Solar (USA, 10%), Suntech Power (China, 7%), Sharp (Japan, 6%), Q-Cells (Germany, 5%).[32] Consolidation of small players and vertical integration is increasing economies of scale and weeding-out the least-efficient production systems. Governments make the market with their generous feed-in-tariffs (FITs). In 2008 and 2009, the industry responded to generous FIT policies and expanded production capacity, but demand declined after Spain revised FITs in 2010, resulting in a surplus of PV

[31]Ibid.

[32]*Renewable Energy and Energy Efficiency Export Initiative*, National Export Initiative (Washington, DC: U.S. Department of Commerce, Trade Promotion Coordinating Committee, 2010), citing 'Renewable Energy Policy Network for the 21st Century, "Renewables 2010 Global Status Report." Available at http://export.gov/reee/.

modules. Factories in the U.S. had to slow production or delay planned expansions.[33]

Over the past ten years, the center of gravity of the global PV industry shifted west and then back east as the industry moved to the countries with the strongest political programs to support domestic construction of solar PV. In 2004, Japan was the leader in cell production (830 MW), followed by Europe (470 MW), China (200 MW), and the United States (150 MW).[34] In 2007 and 2008, Germany produced more solar PV cells and modules than any other country. During this period, China, Taiwan and the Philippines entered the industry as low-cost exporters. Today China is the world's largest producer of PV solar cells, manufacturing a vast 3,800 MW of PV cells in 2009, though it has almost no domestic demand at present.[35] One analysis predicts that "the production of cell, modules, and BOS components will likely continue to shift toward countries in Asia and other low-cost production centers that possess a comparative advantage in flexible light manufacturing of tradable goods."[36]

The price of PV panels decreased over the past decade, despite volatility in the PV supply chain. PV modules consist of many parts, the PV panel, system components and mounting material. The silicon market, a key component of the PV panel, generated much of this this volatility. A pricing oligopoly, with only ten firms supplying silicon and accelerating demand from solar cell producers, increased the spot price of silicon from $30 to $400 per kilogram in 2008, before new capacity created oversupply and reduced prices $60 to $80 per kilogram.[37] The "balance-of-system" components, or all the parts in a

[33] "U.S. Solar Market Insight, 2nd Quarter 2022: Executive Summary" (Washington, DC: Solar Energies Industry Association, 2011). Available at http://www.seia.org/galleries/pdf/SMI-Q2-2011-ES.pdf.

[34] World Bank, *International Trade and Climate Change: Economic, Legal and Institutional Perspectives* (Washington, DC: World Bank Environment and Development, 2007), p. 62.

[35] *Photovoltaic Power Systems Programme (PVPS) Trends Report 2009*, (Paris: International Energy Agency, 2009), p. 25. Available at http://www.iea-pvps.org/index.php?id=32

[36] Jacob Funk Kirkegaard, Thilo Hanemann, Lutz Weischer and Matt Miller, "Toward a Sunny Future? Global Integration in the Solar PV Industry," Working Paper Series 10-6 (Washington, DC: Peterson Institute for International Economics, May 2010), p. 30.

[37] Ibid., p. 18.

solar module other than the PV panel itself, account for 20-70% of the cost. These components include inverters (DC to AC), mounting hardware, cables, batteries, controllers and also non-hardware costs such as shipping, taxes, permitting fees and labor when calculating final installed costs. Balance-of-system costs decreased by the same level as the module between 1998 and 2007, about 40%. However, as module costs continued to decrease, between 2008 and 2009 balance-of-system costs exhibited a slight increase, according to the "Tracking the Sun III" report.[38] Installation of solar PV and CSP is necessarily conducted on site, and naturally dominated by local contractors.

Between 1998 and 2009, manufacturers reduced costs for final, installed PV systems by 30%, about 3.2% annually.[39] Thus, photovoltaic systems are expensive but becoming cheaper, whereas the opposite could be said of fossil-fuel and nuclear generation sources.

Concentrating Solar Power Industry

CSP installed capacity is much smaller than PV, totaling 1 GW in 2010. Aiming to take advantage of supportive policies and new technological advances, developers are planning 15 GW of CSP.[40] CSP reflects the optimism of PV. The IEA's Energy Technology Perspectives 2008 (ETP 2008) forecasts under one scenario that CSP could produce 5% of global electricity production by 2050.[41] Policymakers in the U.S. and EU view CSP as a feasible, cost-effective technology and direct research funding and subsidies towards the sector.

The concentrating solar power (CSP) sector is emergent; 37 CSP plants, at 24 sites, operate today world-wide. Only a handful of companies construct and operate CSP plants, including Spain's Abengoa Solar and Acciona Energy, France's Areva, US firms BrightSource, Luz, eSolar and others. CSP development stalled from the late 1980s

---

[38]Galen Barbose, Naïm Darghouth and Ryan Wiser, *Tracking the Sun III: The Installed Cost of Photovoltaics in the US from 1998-2009* (Berkeley, California: Lawrence Berkeley National Laboratory, Environmental Energy Technologies Division, December 2010).

[39]Ibid., p. 1.

[40]*Technology Roadmap: Concentrating Solar Power* (Paris: International Energy Agency, 2010), p. 9.

[41]Ibid., p. 19.

to 2005. The Saguaro Solar Power Plant in Arizona came online in 2005 after a fifteen year hiatus of any CSP development. Nine new CSP plants have entered operation in the United States since 2005.[42] In Europe, specific feed-in tariff policies (discussed above), resulted in the PS10 CSP plant, the first grid-connect central receiver CSP plant. In Spain, and the United States and some North African countries, governments and companies are planning more than 50 CSP plants.[43]

CSP plants are capital intensive and initial investments costs range from $4.2/W to $8.4/W depending on labor and land costs, technologies, the amount and distribution of solar irradiance and the size of the solar field.

For both PV and CSP, reducing generation costs will require further support to improve technology, cheaper manufacturing and economies of scale. These dynamics draw attention to the central role of socio-political concerns.

Trade Policy for Solar Energy Goods

The solar energy sector developed into a globally traded industry, with countries exporting and importing PV panels, CSP heliostats, generators, thermostats and solar water heaters. As described above, low-cost manufacturers in Asia produce more and more solar equipment. The U.S. and EU subsidize the sector, and consequently face increasing political and popular pressure to protect domestic solar energy production capacity, ensuring that subsidy dollars remain in the domestic economy. These measures could hinder the development of a global solar energy industry and prevent wide-spread deployment of low-carbon solar energy electricity generation.

If anything, since 2007, the U.S. and EU have followed similar strategies, which puts the two economies at odds with each other as they compete for exports. The White House stated in its March 2011 Secure Energy Future strategy that "we must focus on expanding cleaner

---

[42] Utility_Scale Solar Projects in the United States, Operating, Under Construction, or Under Development, website, Solar Energy Industries Association. Accessed October 21, 2011 at http://www.seia.org/galleries/pdf/Major%20Solar%20Projects.pdf.

[43] Arvizu, et. al., *Direct Solar Energy* in: *IPCC Special Report on Renewable Energy Sources and Climate Change Mitigation*, pp. 37–38.

sources of electricity, including renewables like wind and solar, as well as clean coal, natural gas, and nuclear power—keeping America on the cutting edge of clean energy technology so that we can build a 21st century clean energy economy and win the future."[44] The European Commission's March 2, 2010 *Europe 2020 Strategy* stated that "The EU should maintain its lead in the market for green technologies as a means of ensuring resource efficiency throughout the economy, while removing bottlenecks in key network infrastructures, thereby boosting our industrial competitiveness."[45] And the EU's *Roadmap for moving to a competitive low-carbon economy in 2050* stated that "By stepping up climate action 1.5 million additional jobs could be created by 2020."[46] According to a report from Senator Ron Wyden and the Department of Commerce, the U.S. is running a trade deficit in environmental goods (defined as the World Bank 43) of $4.3 billion in 2009. In light of this report, Senator Wyden suggested that the U.S. take action to increase its exports of renewable energy goods and reduce the trade deficit. Both the U.S. and EU aim to export to a limited marketplace, and have focused efforts on reducing non-tariff barriers.

Indeed, tariffs on solar energy goods remain quite low. Though a few countries have tariffs above 10%, the most solar-energy importing markets have no import tariffs on solar power equipment.[47] For the large number of developing countries with tariffs on solar energy goods, the World Bank estimates that eliminating tariffs on environmental goods and services (EGS) in a set of developing countries would increase trade in solar power goods 6.4%. Further eliminating non-tariff barriers would result in a total increase in trade of 13.5%.[48]

---

[44]*Blueprint for a Secure Energy Future*, White House (Washington, DC: The Executive Office of the President, March 30, 2011). Available at http://www.whitehouse.gov/sites/default/files/blueprint_secure_energy_future.pdf.

[45]*Europe 2020: A strategy for smart, sustainable and inclusive growth*, European Commission (Brussels, Belgium: Presidency of the European Commission, March 3, 2010), p. 15. Available at http://eur-lex.europa.eu/LexUriServ/LexUriServ.do?uri=COM:2010:2020:FIN:EN:PDF.

[46]*Roadmap for moving to a low-carbon economy in 2050*, website, European Commission, Directorate-General for Climate Action. http://ec.europa.eu/clima/policies/roadmap/index_en.htm, accessed on October 21, 2011.

[47]Kirkegaard, et. al, op. cit., p. 32.

[48]World Bank, *International Trade and Climate Change: Economic, Legal and Institutional Perspectives* (Washington, DC: World Bank Environment and Development, 2007).

The discussion about facilitation of trade in EGS reached prominence in 2001, when trade ministers at the World Trade Organization's Doha meeting adopted a declaration calling for inter alia "the reduction or as appropriate elimination of tariff and non-tariff barriers to environmental goods and services."[49] Six years later, a group of "Friends of Environmental Goods and Services" developed a grocery list of 153 potential items in 2007, and the World Bank refined this list to 43 environmental goods, suggesting facilitating trade of these goods as a way of easing countries' work towards implementing a low-carbon economy.[50]

The U.S. and EU have followed a cooperative approach in regards to trade in EGS. In an op-ed about the proposed EGS agreement in 2007, the EU Trade Commissioner Peter Mandelson and U.S. Trade Representative Susan Schwab advocated for rapid global adoption of a joint proposal for trade in environmental goods' inclusion in the final Doha trade talks, reducing tariffs on certain environmental goods and services to zero, using the World Bank's list.[51] Subsequently, the international community has stumbled at the level of implementing a system that reduces tariffs on environmental goods. In the turmoil of the global economic recession, progress in this area, and other areas of liberalized trade, has slowed.

In all fairness, it is not a simple lack of political will. Data problems hinder progress. The international trade system of coding goods for the purposes of applying tariffs (HS or SITC system) aggregates related goods, and as result does not differentiate between, for example, mirrors used for CSP or mirrors for other industrial use. Likewise, photovoltaic panels fit into the same category as Light-Emitting Diodes (LEDs), many which are not used as environmental goods. Data for the materials and goods used in the construction of CSP is

---

[49] *Doha Ministerial Declaration* WT/MIN(01)/DEC/1, World Trade Organization (Doha, Qatar: WTO, November 14, 2001).

[50] World Bank. *International Trade and Climate Change: Economic, Legal and Institutional Perspectives*. Washington, DC: World Bank Environment and Development, 2007), p. 79.

[51] Susan Schwab (U.S. Trade Representative) and Peter Mandelson (EU Trade Commissioner), "Working towards an open global market in green technology," *Wall Street Journal*, December 7, 2007.

much more difficult to aggregate, because the harmonized system of custom codes does not differentiate between the final use of the mirror or pipes that carry the steam.[52] Countries, especially developing countries that collect significant customs revenues, hesitate to reduce or eliminate tariffs on an overly-large group of goods just to attempt to increase trade in EGS. Therefore, EU and U.S. efforts have targeted non-tariff barriers.

One example of the non-tariff barriers that the EU and U.S. are confronting are domestic content requirements (DCR) or local content requirements (LCR). In August 2011, the EU filed a WTO complaint against the Canadian province of Ottawa for its DCR for solar and wind projects. In October 2011, U.S. solar companies, alleging dumping, asked the U.S. government to consider placing import tariffs of 100% Chinese solar panels. These actions both indicate commitment to maintain liberalized trade, but also the potential for conflict between the EU and U.S. as they seek to increase their international market share in the face of growing competition from Asia.

The second area of trade is the actual export of electricity that is generated from solar energy. Unlike oil or gas, which can be contained and piped, electricity is fugitive and expensive to store or transmit. In a few cases, such as the transmission of hydropower from Canadian hydroelectric dams to the United States, "renewable electricity" itself is actually the traded commodity. This area of solar energy trade is especially important for CSP, which is most effective in desert areas, far from population centers. Many industrialists envision the generation of electricity in areas with clear skies and high solar irradiance, such as North Africa, Australia or Southwestern United States, and the transmission of this electricity across long-distance high-voltage lines. For example, North African countries could directly deliver power to the European Union, Australia to Indonesia, or Arizona to California.

[52] Two papers discuss the problems with identifying environmental goods and services: Senator Ron Wyden, "US Trade in Environmental Goods. Follow-Up Report To Major Opportunities And Challenges To US Exports Of Environmental Goods," (Washington, DC: Office of Senator Ron Wyden. US Senate, May 20, 2010), p. 3; Kirkegaard, et. al, op. cit., p. 28.

DESERTEC

The DESERTEC initiative demonstrates many of the issues discussed in this chapter. DESERTEC aims to generate electricity for Africa and Europe from the installation of CSP, PV and wind facilities across the Middle East and North Africa (MENA).

The DESERTEC Foundation and Industrial Initiative is promoting the establishment of giant solar fields in the Sahara Desert that will be able to supply about 17% of Europe's electricity usage by 2050. The Industrial Initiative has attracted 17 shareholders from eight countries, including many of the major players. Many other European and international organizations are looking at the southern coast of the Mediterranean as one huge green electricity generation site. The World Bank's project on concentrated solar power in the MENA region aims to co-finance commercial-scale power plants to help MENA countries become suppliers and consumer of solar-generated electricity. The EU's Union of the Mediterranean Solar energy plan, while quieter in the publicity front, is also working to develop projects in Morocco and Tunisia.

An examination of the demographic and energy data for the MENA region should cause reason for doubt. There are technological challenges—cost-competitive generation, maintenance and transmission—and political challenges, as governments across North Africa undergo difficult transitions. The largest challenges arise from growing populations and their related increasing demands for energy, which will utilize the full capacity of any electrification projects, and the inefficient use of energy resources in North Africa.

The demand for electricity exists. In North Africa alone, the demand challenge is chilling. The youth cohort is the largest in the world, and increasing development has increased electricity demands. Oil and gas, which North Africa exports, is now being used domestically as industry develops. The European Union anticipates increasing energy imports from North Africa for a different reason: it wants to obtain 'green energy.' The EU's "20-20-20" energy package provides a legislative requirement for the use of renewables and targets for member countries to reduce their overall greenhouse gas emissions. European countries have limited capacity for domestic renewable energy generation, but the desert of North Africa provides a rich "Sun Belt" for concentrating solar power (CSP) and sustained winds for aeolian power.

Beyond the technological challenges to design and build the projects, the successful implementation of a solar energy system across North Africa is dependent on two initial steps. First, the industrial groups working in North Africa must collectively agree that the top priority is providing energy for North Africans. This is a developmental goal, and also a regional requirement. Second, on the demand side, North African countries must increase energy efficiency. While all other regions decreased their energy use per $1,000 GDP, in North Africa energy use has increased more than 100% in the past twenty years. If these initial steps are taken, solar-generated electricity can provide for North Africa's development while also benefitting the EU market.

Sources: DESERTEC Foundation. Website. http://www.desertec.org/. World Development Indicators, from International Energy Agency (IEA Statistics © OECD/IEA, http://www.iea.org/stats/index.asp) and United Nations, Energy Statistics Yearbook.

The Way Forward

The European Union and the United States followed remarkably similar paths towards the development of solar energy policy, regardless of substantial differences in the details. Both countries face the problem of competition from countries with lower labor costs and environmental standards. The U.S. and EU could use this area to reinvigorate cooperation and indicate that the transatlantic relationship today is about more than simply stopping financial market contagion; it is also about taking on constructive projects.

First, the U.S. and EU should maintain the momentum of the current wave of solar energy innovation. This requires the political assuredness to continue costly front-end financing of research and prototype development so that nascent solar energy technologies cross the 'innovation valley of death'. A joint effort, possibly as the marquee initiative of the EU-U.S. Transatlantic Energy Council, could provide both with the mutual confidence. It requires even more political assuredness to develop long-term feasible market measures. National feed-in-tariff policies provide flawed incentives for the development of solar energy, as demonstrated in Spain, but they are the best option when well-designed. Work, such as the U.S. National Renewable Laboratory's "Policymaker's Guide to Feed-in Tariff Policy Design" should be continued at the transatlantic level to enhance systems. The United States especially has lost the resolve for climate change measures, such as Renewable Portfolio Standards, or a carbon system, so a transatlantic market-based measure is unlikely.

Therefore, as a second step, to build a global market, the EU must forge ties with Asia to maintain the growth of PV and CSP deployment. These regions have the largest current installations of solar power capacity, and with China's ambitions to build a domestic market in addition to exporting solar equipment, the EU-Asia partnership will be the future solar market.

Thirdly, the U.S. and EU must renew the stalled efforts to reduce tariffs and non-tariff barriers on environmental goods and services. The U.S. and EU will continue to maintain the lead in the development and innovation of solar energy technologies. Corporations based in the U.S. and EU will lead in the design and installation of systems, profiting from upstream innovative activity, as well as down-stream local activity.

Despite the rise and fall of previous waves of solar energy innovation, this moment is unique. Solar energy technology is mature. Systems in operation today come tantalizing close to cost recovery without subsidies. Countries, politicians and societies desire renewable energy, including solar, for many mixed motives. Current efforts recall the optimism about solar power from earlier in this decade, when researchers and policymakers believed that solar power could help poor countries develop and mitigate greenhouse gas emissions. The case for solar energy is stronger than ever.

Chapter Eight

Transatlantic Cooperation for a Competitive and Sustainable Biofuel Industry

Tamás Kenessey

Humanity has never before faced so much uncertainty about the sustained availability of resources and energy indispensable for its future development. The *International Energy Agency's World Energy Outlook 2010* projects that world primary energy demand will increase by 36 percent until 2035, even if nations implement recently adopted policy measures on resource efficiency. Fossil fuel use is restricted by many factors, such as the increasing costs of extraction, the situation of reserves in politically unstable parts of the world, the need to reduce anthropogenic greenhouse gas (GHG) emissions and the energy importing nations' pursuit of greater energy security. In order to limit the global temperature rise to below 2°C governments have to take bold actions without delay. GHG emissions should be cut by at least one-third compared with business-as-usual models until 2050. Renewable energy will play a central role in the future energy mix. In the short term, renewables are required to meet the world's increasing demand in a sustainable and economically competitive way. In the medium- and long term, their role is to gradually replace fossil fuels. According to the special report of the United Nations Intergovernmental Panel on Climate Change (UN IPCC), with the right backing from policymakers they could provide up to 80 percent of the world's primary energy supply by 2050.[1]

[1] O. Edenhofer, R. Pichs, Madruga, Y. Sokona, K. Seyboth, P. Matschoss, S. Kadner, T. Zwickel, P. Eickemeier, G. Hansen, S. Schlömer, C. von Stechow, ed. "IPCC, Summary for Policymaker," in *IPCC Special Report on Renewable Energy Sources and Climate Change Mitigation* (Cambridge, United Kingdom: Cambridge University Press, 2011). Available at: http://srren.ipcc-wg3.de/report.

Definition of Biofuels

Biofuels are liquid or gaseous fuels produced from organic matter derived from plants, animals and waste and primarily used in the transport sector. They can be classified according to feedstock, technological pathways, historical development or the maturity and commercialization of the technology. *First generation* biofuels are based on traditional feedstock such as sugarcane, corn, soybeans and palm oil. These technologies have already reached technological maturity and production is done on commercial scale. *Second generation* biofuels use agricultural and forestry residues and left-overs and usually apply cellulosic conversion technologies. *Third generation* biofuels are produced from a wide range of feedstock specially designed for energy purposes and apply advanced conversion technologies that are currently under development. Another classification divides biofuels to *conventional technologies* that use well-established processes on industrial scale and *advanced technologies* that are still in the R&D or demonstration stage and need further improvement before commercialization.

Biofuels in the Global Energy Mix

Biofuels will play a central role in gradually replacing fossil fuels in the transport sector. The International Energy Agency (IEA) estimates that the transport sector accounts for half of the global primary oil consumption. Transport is almost entirely based on petroleum which supplies 95 percent of its total energy use. At the same time, according to conservative assessments, transport was responsible for approximately 23 percent of all anthropogenic GHG emissions in 2004 and since then this share has increased.[2] The IEA has estimated that transport related emissions could double between 2000 and 2050 and the bulk of these will occur in non-OECD countries.[3] Biofuels currently

[2] B. Metz, O.R. Davidson, P.R. Bosch, R. Dave, L.A. Meyer , ed., "Transport and its infrastructure," in *Climate Change 2007: Mitigation. Contribution of Working Group III to the Fourth Assessment Report of the Intergovernmental Panel on Climate Change* (Cambridge, United Kingdom and New York, NY, USA: Cambridge University Press, 2007). Available at: http://www.ipcc.ch/pdf/assessment-report/ar4/wg3/ar4-wg3-chapter5.pdf.

[3] *World Energy Outlook 2010* (Paris, France: OECD / International Energy Agency, 2011).

provide around three percent of total transport fuel and their share is quickly increasing. According to the *IEA's Technology Roadmap Biofuels for Transport* the share of biofuels can grow to 27 percent of total transport fuel by 2050, saving 2.1 gigatonnes of CO_2 emissions yearly.[4] To meet this ambitious target, a number of challenges have to be overcome. Economic and sustainable production of feedstock, effective conversion pathways and the compatibility of the final product with current distribution and end-of-use infrastructure are the main development bottlenecks. A number of serious *social and ethical issues* also have to be addressed. Evidence showed that conventional biofuels made from foodstuffs contributed to soaring food prices in 2007. Due to the indirect land use change effect of feedstock production, their greenhouse gas balance proved to be worse than previously expected. But for all their present flaws biofuels have a huge potential. They can reduce our dependency on oil and consequently improve energy security. Advanced technological pathways using waste, agricultural and forestry residues or plants specially designed for energy use offer substantial GHG reductions and avoid the food versus fuel conundrum. Partly owing to recent events in the energy industry, partly to economic considerations and partly to developments in the biofuel industry, the ground is shifting back towards biofuels.

The Biofuel Industry Today

The global biofuel production in 2010 exceeded 100 billion liters and provided nearly three percent of global transport fuel supply. The production increased more than fivefold in the last ten years. Continued policy support, high oil prices in recent years, the need to reduce GHG emissions in the transport sector and technological innovations all contributed to the rapid expansion. Forecasts say that demand will continue to rise in the foreseeable future and biofuels can provide one-tenth in 2030 and up to 27 percent in 2050 of all transport fuels.[5] Cur-

[4] *Technology Roadmap, Biofuels for Transport* (Paris, France: OECD / International Energy Agency, 2011).

[5] The calculations derive from the *BP Statistical Review of World Energy* and the IEA's Technology Roadmap, Biofuels for Transport. *BP Statistical Review of World Energy*, (London, United Kingdom: BP p.l.c., 2010) and *Technology Roadmap, Biofuels for Transport* (Paris, France: OECD / International Energy Agency, 2011).

Figure 1. Share of U.S., Brazil, EU and rest of the world in bioethanol (2009)

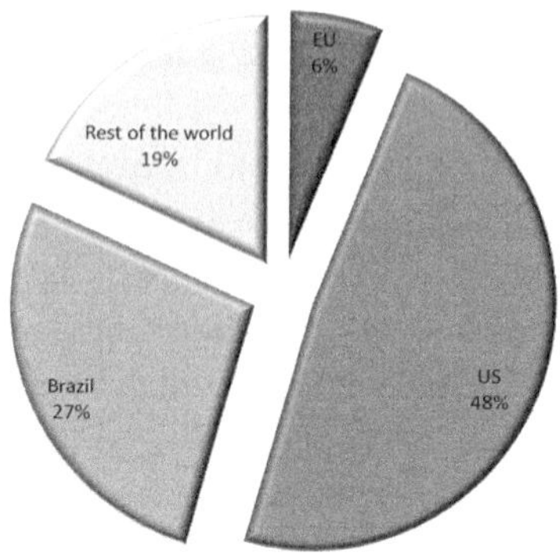

Source: World Energy Outlook (2010) and Technology Roadmap: Biofuels for Transport (2011), OECD / International Energy Agency

Figure 2. Share of EU, U.S. and rest of the world in biodiesel (2009)

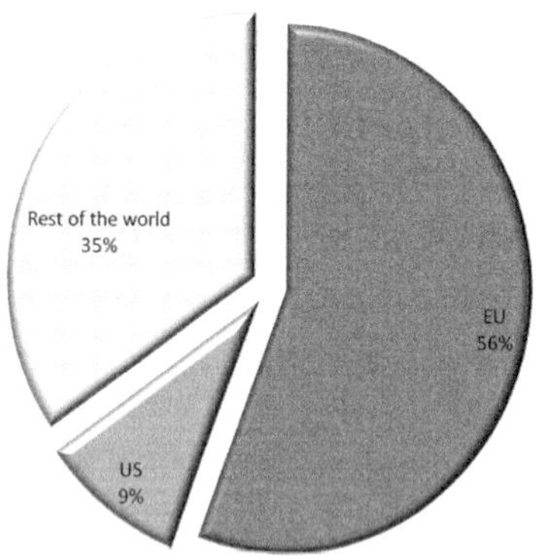

Source: World Energy Outlook (2010) and Technology Roadmap: Biofuels for Transport (2011), OECD / International Energy Agency.

rently the United States, Brazil and the European Union dominate the market. They all apply a mix of policy measures including R&D and investment support, blending mandates, tax incentives and certification schemes to stimulate the growth and development of the industry.

Brazil

In many respects, the South American nation can provide an example for other countries on how to build a market-based and sustainable biofuels industry. Brazil has been in the forefront of ethanol production since the 1970s. As a result of continuous state support, long-term targets and a relatively stable regulatory environment, companies made significant investments in the sugarcane ethanol sector. Since the 1990s the government gradually reduced ethanol subsidies and abolished production quotas. Today the price of sugar and ethanol is regulated by the market and this led to substantial efficiency gains. Brazil currently applies a 20-25 percent blending mandate for bioethanol and a five percent blending mandate for biodiesel as per 2013. Sugarcane-based ethanol provided 21 percent of Brazil's road transport fuel demand in 2008. Brazilian sugarcane ethanol is a conventional technology, but its price is competitive with petrol and it offers up to 86 percent GHG reduction compared to petrol.[6]

United States

Corn-based biofuel production started in the United States following the 1973 oil crisis. With almost 50 billion liters in 2010, the United States is currently the world's leading producer of bio-ethanol and has an ambitious target to produce 136 billion liters by 2022. 60 billion liters of this should derive from advanced, lingo-cellulosic feedstock. Biodiesel production is not significant compared to ethanol.[7] Corn-based ethanol production is relatively cheap, but serious concerns had been raised about its sustainability. Almost 40 percent of all U.S. corn is

---

[6] Searchinger et al. (2008) in Perrihan Al-Riffai, Betina Dimaranan, David Laborde, *Global Trade and Environmental Impact Study of the EU Biofuels Mandate* (Brussels, Belgium: International Food Policy Institute / European Commission, 2010).

[7] Production of 2 billion liters in 2009. Data source: National Biodiesel Board http://www.biodiesel.org/.

expected to go to biofuel production in 2011,[8] and among all biofuels U.S. corn-based bioethanol is believed to have had the greatest effect on soaring food prices in 2007. Moreover, corn production requires fossil fuels, a great quantity of fertilizers, and water. The U.S. applies a complex system of support schemes. On the federal level this includes tax reductions, direct subsidies for ethanol and biodiesel, and a tariff system designed to limit imports especially from Brazil. States are also allowed to adopt tax exemptions, subsidies and blending mandates different from federal ones.[9] In June 2011 a cross-party movement in the U.S. Senate voted to end the corn-based ethanol tax credits and the 54 cent per gallon tariff on ethanol imports. The bill, however, extended tax credits for the cellulosic ethanol industry until 2015.[10] Capital investments also receive subsidy from the government. President Barack Obama unfolded the administration's new energy security plan in March 2011.[11] The president announced to spend up to 510 million USD to build four new bio refineries with a combined capacity of more than 80 million gallons (approximately 300 million liters), and offered a special partnership for the private sector with the U.S. Department of Energy, Department of Agriculture, the U.S. Navy and U.S. Air Force. Government support is needed as technical difficulties slowed down the development and commercialization of advanced biofuels. Due to the higher than expected production costs, the U.S. Department of Agriculture had to scale down its estimated capacity of advanced biofuels to only 40 million liters, one-tenth of the originally expected amount for 2010.

---

[8] *Feed Grains Data, August 2011*, United States Department of Agriculture, Economic Research Service. Available at: http://www.ers.usda.gov/Data/FeedGrains/FeedYearbook.aspx#FSI.

[9] Perrihan Al-Riffai, Betina Dimaranan, David Laborde, *Global Trade and Environmental Impact Study of the EU Biofuels Mandate* (Brussels, Belgium: International Food Policy Institute / European Commission, 2010).

[10] "Senate deal would axe $6 billion ethanol tax credit," *Reuters*, July 7, 2011. Available at: http://www.reuters.com/article/2011/07/07/us-usa-ethanol-deal-idUSTRE7663OS20110707.

[11] *Remarks by the President on America's Energy Security*, March 30, 2011. Available at: http://www.whitehouse.gov/the-press-office/2011/03/30/remarks-president-americas-energy-security.

The European Union

Biofuels provide approximately four percent of all transport fuels in the EU27. The EU's biofuel industry is dominated by biodiesel production, owing to the fact that diesel engines account for roughly half of the European car market. In 2009 European biodiesel production exceeded 10 billion liters, around 56 percent of world production.[12] It is entirely based on first generation feedstock, such as rapeseed, sunflower and soybean. Bioethanol production from corn and wheat was 3.7 billion liters in 2009.[13] The European Union has a target of 10 percent renewable energy in transport by 2020, for which lignocellulosic biofuels and biofuels made from waste and residues count twice. To help achieving this target the EU authorized member states in an Energy Tax Directive to introduce tax reductions for biofuels. The EU's Common Agricultural Policy since 2003 supports the production of energy crops in the form of decoupled direct payments and targeted support. The EU seemed to be on track to meet its 10 percent target, but the recent debate on the indirect land use change (iLUC) effect of European biofuel policies and the lack of commercially viable advanced technologies slowed down the progress. Another restrictive factor is that the biomass producing capacity of Europe is already exploited to 40-75 percent.[14] Compared to other parts of the world, the EU is short of unused arable land and the future growth of European biofuel industry will need feedstock import. As a consequence the EU will be interested in gradually dismantling the trade barriers that currently limit the free flow of feedstock and biofuels on world markets.

---

[12]Source of data: Biofuels platform—http://www.biofuels-platform.ch/en/infos/eu-biodiesel.php

[13]Ibid.

[14]M. Altmann, P. Schmidt, W. Weindorf, Z. Matra, A. Brenninkmeijer, J.-C. Lanoix, O. van den Kerckhove, C. Egenhofer, A. Behrens, J. Nuñez Ferrer, R. Bleischwitz, A. Crisan, *The assessment of potential and promotion of new generation of renewable technologies* (Brussels, Belgium, study commissioned by the European Parliament, 2011). Available at: http://www.lbst.de/ressources/docs2010/EP-02_Renewables_JUNE 2010_PE-440-278.pdf.

Conventional Biofuels and Their Shortcomings

Sugar and starch-based conventional bioethanol currently provides more than three-quarters of all biofuel production. It has been a hundred years since Henry Ford designed the famous T-model to run on this fuel. The basis of the process is the fermentation of sucrose or glucose to bioethanol. Corn, sugarcane, sugar beet, sweet sorghum and wheat are the most frequently applied feedstock. Conventional biodiesel can be produced from vegetable oils, animal fats and used cooking oil by a chemical process called transesterification. Mostly sunflower, rapeseed, soybean and palm oil are used as feedstock. Biogas is the result of the anaerobic digestion of organic waste, animal manure, sewage sludge and plant residues. It can be directly burnt to heat and produce electricity. Cleaned from contaminations and upgraded to biomethane it also can be injected into natural gas grids or used in natural gas vehicles. The share of biogas in transport fuels is currently negligible. The production costs of conventional biofuels are already comparable to that of fossil fuels. The IEA estimates that corn and cane bioethanol is produced for 62-75 U.S. cents per liter gasoline equivalent,[15] depending on yield, geographical and weather conditions. At the same time, in 2010 average prices, a liter of petrol is produced for about 54 cents.[16]

Food vs. Fuel

Although they have become increasingly competitive in recent years, conventional bioethanol and biodiesel have a number of shortcomings. They are produced from the edible part of plants and their production relies heavily on fossil fuels. Consequently they contributed in recent years to the increase in food prices and had little effect to cut greenhouse gas emissions. Measuring the impact of biofuels on food prices is a complex task. They played an undeniable role in the 2007-2008 food price crisis together with other factors, such as

[15] Liter gasoline equivalent is the amount of alternative fuel it takes to equal the energy content of one liter of gasoline (energy content 33.5 MJ/liter).

[16] Costs are calculated based on global average retail price without taxation. Source of data: *Technology Roadmap, Biofuels for Transport*, (Paris, France: OECD / International Energy Agency, 2011).

the increased demand for meat products in emerging economies, adverse or extreme weather conditions in major producer countries, low level of global stocks, protective policies on the export or import of foodstuffs in certain countries and, last but not least, increased oil prices. Sophisticated global equilibrium models came to the conclusion that "although individual crop prices appear to be affected by biofuels, the impact of biofuels on global or aggregated food prices is rather small."[17] An extensive 24 month-long study found that only 12 percent of the rise in the IMF's food price index could be attributed to biofuels.[18] Approaching the problem from the other end, however, shows that rising crop prices seriously affect the competitiveness of conventional biofuels. The price of the final product *depends heavily on feedstock prices*, which represent 45 to 70 percent of production costs.[19] The price of sugarcane, corn and wheat became ever more volatile in recent years. The mounting demand for food, fiber and energy crops of the world's growing population will result in even higher agricultural prices. Future feedstock production is also dependent on available arable land, and additional intensive inputs like water and fertilizers. The U.S. Department of Agriculture predicts that if the United States 2030 biofuel target is met using only corn-derived ethanol, agricultural water use could increase six-fold.[20] Land, water and fertilizer demand and the consequent competition with food production are the most serious economic constrains of conventional biofuel technologies.

---

[17]Govinda R. Timislina & Ashish Shreshta, Environment and Energy, Development Research Group, *Biofuels: Markets, Targets and Impacts*, (Washington DC, United States: The World Bank, 2010).

[18]Baier, Scott, Mark Clements, Charles Griffiths, and Jane Ihrig, "Biofuels Impact on Crop and Food Prices: Using an Interactive Spreadsheet," in *International Finance Discussion Papers*: Number 967 (Washington DC, United States: Board of Governors of the Federal Reserve System, 2009).

[19]*Transport, Energy and CO_2* (Paris, France: OECD / International Energy Agency, 2011).

[20]K.C. Stone, P.G. Hunt, K.B. Cantrell, K.S. Ro, *The potential impacts of biomass feedstock production on water resource availability* (Washington DC, United States: United States Department of Agriculture, Agricultural Research Service, 2009). Available at: http://ddr.nal.usda.gov/bitstream/10113/38481/1/IND44306549.pdf.

Table 1. Greenhouse Gas Balances of Biofuels in 2005

GHG Emissions (% change)

	Feedstock	Previous Land Use	No LUC	With Direct LUC	With indirect LUC
Biodiesel	Waste oil	n.a	-90	n.a	n.a
	Rapeseed	Cropland	-58	-58	+5 to + 69
	Rapeseed	Pasture	-58	-25	+39 to + 102
Ethanol	Sugar cane (Brazil)	Cropland	-71	-71	-35 to +1
	Maize	Cropland	-55	-55	-22 to + 11
	Maize	Pasture	-55	-37	-5 to + 28
	Wheat	Cropland	-49	-49	+6 to + 63
	Wheat	Pasture	-49	-22	+ 36 to + 92

Source: Fritsche & Wiegmann (2008) in Fischer et al. (2009)

Note: Waste oil includes waste vegetable and animal oils. LUC stands for land use change. Direct LUC refers to emissions including those arising from the land conversion to cultivate the biofuel, whereas indirect LUC refers to emissions including those arising from the conversion of land elsewhere to replace production displaced by biofuel cultivation.

Greenhouse Gas Savings

The greenhouse gas balance of conventional biofuels also failed to meet expectations. *Fritsche & Wiegmann (2008)* estimated that most of the conventional feedstock and conversion technologies result in additional GHG emissions compared to fossil fuels if direct and indirect land use change is included.[21]

Four studies leaked from the services of the European Commission in July 2011 to the Reuters news agency came to similar conclusions. They suggest that the EU's renewable energy target in the transport sector may lead to additional emissions instead of avoiding greenhouse gases. The European Commission's internal calculations state that compared to the CO_2 emissions of 83.8 grams per megajoule for fossil fuel, palm oil causes 105 grams, soybean 103 grams, rapeseed 95 grams, sunflower 86 grams, wheat 47 to 64 grams, corn 43 grams of CO_2. Only sugar beet's 34 grams, sugarcane's 36 grams

[21]Fritsche & Wiegmann (2008) in Fischer, Günther, Eva Hizsnyik, Sylvia Prieler, Mahendra Shah and Harrij van Velthuizen, *Biofuels and Food Security* (Vienna, Austria, prepared by the International Institute for Applied Systems Analysis for OPEC Fund for International Development, 2009).

and cellulosic ethanol's 9 grams of CO_2 emissions result in substantial GHG savings.[22]

Energy Density

A further weakness is that the energy density of bioethanol is only about 65 percent of that of gasoline, while biodiesel's energy performance is around 90 percent of diesel fuel's.[23] Thus, the blending of conventional biofuels is limited to 10-15 percent in the case of ethanol and around 20 percent for biodiesel. Above this so-called *"blending-wall"* only flex-fuel cars with modified engines can run on higher biofuel blends.

Future Prospects for Biofuels

For all their present day weaknesses, biofuels are considered to be an integral part of the future energy mix in the transport sector. Biofuels can reduce our dependency on oil, improve energy security and cut back greenhouse gas emissions. They are energy rich, liquid, easy to transport and almost entirely compatible with the current distribution and end-of-use infrastructure. Until the proliferation of plug-in hybrid electric vehicles (PHEVs), which require the dramatic development of battery technology and the considerable improvement of the current electric network, there is no real alternative to biofuels in the transport sector. Even with PHEVs on the market, diesel and kerosene replacements are expected to gain further ground in the heavy transport modes and air transport that have limited low-carbon fuel alternatives. The IEA predicts that by 2050 biofuel demand will reach 32 exojoules (EJ), up from today's 2,5 EJ, resulting in more than 100 million hectares in land demand.[24] Biofuel's share in total transport fuels will increase from approximately three percent to around 27 percent.

---

[22] "Factbox: What EU studies say on biofuels' indirect damage," *Reuters*, Brussels, Jul 8, 2011. Available at: http://uk.reuters.com/article/2011/07/08/us-eu-biofuel-factbox-idUKTRE7672XF20110708.

[23] *Bioenergy Conversion Factors* (Oak Ridge, United States: Oak Ridge National Laboratory, 2008).

[24] *BLUE Map Scenario—Energy Technology Perspectives, Scenarios & Strategies to 2050*, (Paris, France OECD / International Energy Agency, 2010).

Biomass is a plentiful resource of energy. The maximum global technical biomass potential from sources that can be sustainably exploited is estimated at 475 EJ by 2050.[25] This is almost triple of the bioenergy (including biofuels) demand projected in the *BLUE Map Scenario of the IEA by 2050.*[26] To harness that amount of energy, however, substantial investments and infrastructural developments have to be made, and the IEA estimates that an additional 70 million hectares of land will be necessary to meet this target.

Conventional technologies will continue to improve in efficiency and will be increasingly competitive with petrol. Their environmental performance will also get better. The IEA predicts that the majority of growth until 2020 will be met by conventional biofuels. Subsequently, however, advanced biofuels are expected to reach technological maturity and have to be deployed on a commercial scale. New technological pathways have to be developed that can use a broad range of feedstock and transform biomass into fuel in a cost effective and environmentally sustainable way. There is no one-size-fits-all solution; at least a dozen of promising technological pathways compete for commercialization. The future of the biofuel industry lies in the diversity of feedstock, conversion technologies and final products. This diversity makes biofuels capable of adjusting to the needs of different geographic regions and transport sectors.

Advanced Technological Pathways

Petrol sets a high standard for the biofuel industry. Ideal biofuels should have similar characteristics as the fossil fuel they are intended to substitute. They have to store as much energy as gasoline or diesel. They have to be compatible with current processing chains in existing refineries, with distribution networks or directly with the vehicles that

---

[25]Dornburg, V., Faaij, A., Langeveld, H., van de Ven, G., Wester, F., van Keulen, H., van Diepen, K., Ros, J., van Vuuren, D., van den Born, G.J., van Oorschot, M., Smout, F., Aiking, H., Londo, M., Mozaffarian, H., Smekens, K., Meeusen, M., Banse, M., Lysen E. andS. van Egmond (2008), *Biomass Assessment: Assessment of global biomass potentials and their links to food, water, biodiversity, energy demand and economy*, MNP, Bilthoven.

[26]*Energy Technology Perspectives, Scenarios & Strategies to 2050*, (Paris, France: OECD / International Energy Agency, 2010).

use them. They should go beyond the food versus fuel debate by using a broad range of non-edible biomass that can be grown on marginal land with relatively little amount of agricultural inputs. Finally, they should provide considerable GHG reductions compared to fossil fuels. The next part presents the most promising technological pathways and the technological challenges they have to overcome in order to reach technological maturity and commercial scale.

Lignocellulosic Bioethanol

Bioethanol can be produced by breaking down the entire lignocellulosic structure of the plant instead of just using its sugar or starch rich parts.[27] The process requires pretreatment using heat and strong chemicals, then enzymes or microbes to liberate sugars. Sugars are then fermented and distilled to produce ethanol. The technological challenge is to find the ideal biochemical process that breaks down the complex polymers to simple sugars in a cost effective way. Different processes use a broad range of feedstock from agricultural and forestry residues like corn stalks and wood chips to energy crops. The ideal energy crop is fast growing, perennial, can be cultivated on marginal land and has little water requirements. *Willow* and *poplar trees*, perennial grasses such as *miscanthus* and *switchgrass* are considered to be the best candidates for the future feedstock of advanced cellulosic bioethanol. Cellulosic ethanol has better GHG balance and performs better in terms of land use requirements than conventional technologies. The IEA calculates that the average cost of a liter of petrol equivalent to cellulose-based ethanol is around 1.1 U.S. dollar. Although this production cost is almost double of conventional bioethanol's, demonstration plants producing cellulosic ethanol are already running on both sides of the Atlantic. *Commercial deployment* has also started in 2011.

Next Generation Biodiesels

Heavy-duty vehicles and aviation have an especially big stake in high-energy content replacements of diesel and kerosene-type fuels. As the IEA put it, "Advanced biodiesel and bio-kerosene will become

[27]Plant cell walls of woody biomass contain three types of carbon-based polymers: cellulose, hemicelluloses and lignin.

increasingly important (...) since demand for low-carbon fuels with high energy density is expected to increase significantly in the long term."[28] Advanced biodiesel can be produced either by *hydrogenating vegetable oils*,[29] or with a two step process in which biomass is first gasified and then converted with the *Fischer-Tropsch synthesis* to a high energy density synthetic liquid (bio-oil). Advanced biodiesel is currently in the *pilot and demonstration stage* and could become fully commercialized in the near future.

Bio-Synthetic Gas

Biomethane (syngas) can be produced via the gasification or pyrolysis of biomass.[30] After purification, the *syngas* can be injected in natural gas networks, in natural gas vehicles, or can be converted to liquid fuel using the Fischer-Tropsch synthesis. The advantage of this technology is that it can use a broad range of feedstock.

Algae-Derived Biodiesel

For a long time the most anticipated technological pathway has been the algae-based biofuel production. Algae can yield up to one hundred times more oil per hectare than conventional sources like soybean or sunflower. Production can take place in artificial open ponds or in so-called closed bio-reactors on non-arable land. Certain algae-strains are able to use brackish, salt or wastewater, hence reducing the fresh water use. They reproduce quickly and yield high energy density oils and by-products rich in carbohydrates. Algae-based biofuel production, however, is still in the *R&D and demonstration stage*. The estimated cost of the raw oil from algae ranges from 0.75 to 5 U.S. dollars per liter.[31] In its 2009 report Accenture states that "algae will be the most difficult and will take the longest to achieve commer-

[28] *Technology Roadmap, Biofuels for Transport* (Paris, France: OECD / International Energy Agency, 2011).

[29] Hydrotreated vegetable oil (HVO).

[30] Gasification is process that converts biomass to a mixture of gaseous materials (syngas) at high temperatures, with a controlled amount of oxygen. Pyrolysis occurs under high pressure and high temperature in absence of oxygen.

[31] *Technology Roadmap, Biofuels for Transport*, (Paris, France: OECD / International Energy Agency, 2011).

Table 2. Oil yields for algae and other biodiesel feedstock

Feedstocks	Oil yield (barrels/ha/year)
Soybean	2.5
Sunflower	5
Jatropha	12
Palm oil	36
Algae	360

Source: Claude Mandil and Adnan Shihab-Eldin, Assessment of Biofuels Potential and Limitations, a report commissioned by the International Energy Forum, February 2010.*

*Another calculation came to the conclusion that algae can yield 1200 gallons (approximately 4540 liters) of oil per acre compared to a yield of 48 gallons (approximately 181.5 liters) from soybean. Source: Darzins, Al, Sustainable Algal Biofuels at Scale: A Prospectus (Albuquerque, New Mexico, U.S., 2009), paper presented at the annual Southwestern Biofuels Policy Summit, May 27-28, 2009.

cial scale."[32] Algae are considered essential for the production of high energy density biodiesel, airline drop-ins and jet fuel on scale. For their promising high productivity and their energy performance several big oil companies made considerable investments in algal biofuel research and pilot plants.

Other Biofuel Technologies

To overcome the problem that ethanol's low energy density presents, scientists also research biochemical processes that result in longer hydrocarbon chains. *Butanol* packs almost as much energy as gasoline and can be blended with fuel in higher proportions. The technological challenge here is to genetically manipulate yeast strains to produce butanol instead of ethanol from sugars. The first butanol-producing commercial plants are expected to open in 2012-2013.[33] *Dimethylether* (DME) is a promising fuel in diesel engines and can be produced from syngas. But the technology is still in its demonstration stage. *Pyrolysis oil*, or bio-crude, can be produced by rapidly heating the biomass to high temperatures and then cooling it down. Bio-crude

[32]M. Stark et al., *Betting on Science: Disruptive Technologies in Transport Fuels*, (Accenture, 2009).

[33]Neil Savage, "The Ideal Biofuel," *Nature* Vol. 474, S9–S11, 23 June 2011.

Table 3. Commercialization status of main biofuel technologies

	Advanced biofuels			Conventional biofuels
	Basic and applied R&D	Demonstration	Early commercial	Commercial
Bioethanol		Cellulosic ethanol		Ethanol from sugar and starch crops
Diesel-type	Biodiesel from microalgae; Sugar-based hydrocarbons	Btl-diesel (from gasification or Fischer-Tropsch synthesis)	Hydrotreated vegetable oil	Biodiesel (by transesterification)
Other fuels and additives	Novel fuels (e.g. furanics)	Biobutanol, DME, Pyrolysis-based fuels		
Biomethane		Bio-synthetic gas	Methanol	Biogas (anaerobic digestion)
Hydrogen	All other novel routes	Gasification with reforming	Biogas reforming	

Source: *Biofuels for Transport, Technology Roadmap,* OECD / International Energy Agency, Paris, France, 2011

is compatible with existing refining and distribution networks and can be processed similar to petroleum[34].

In its 2009 report[35] focusing on the technological development of the biofuel industry Accenture projects the commercial availability of cellulosic bioethanol, butanol and advanced biodiesel in five years. Dedicated energy crops developed with the support of genetic engineering, agricultural, forestry and municipal solid waste is expected to be the dominant feedstock in five to ten years. Bio-crude and complex bio refineries that process biomass feedstock in a wide-range of fuels and by-products should be launched on commercial scale at the same

[34]A more detailed description of these and other additional technologies is available in the *Technology Roadmap, Biofuels for Transport*, in M. Stark et al., *Betting on Science: Disruptive Technologies in Transport Fuels* (Accenture, 2009) and in Neil Savage, "The Ideal Biofuel" *Nature* Vol. 474, S9–S11, 23 June 2011.

[35]Ibid.

time. Algae-based biofuels and jet-fuels are predicted to reach commercial scale in the next 15 years.

The Role of Policymakers

Technological progress is essential, but without the right regulatory support biofuels will not be able to fulfill their role in decarbonizing the transport sector. Policymakers should not only focus on supporting the R&D efforts or adopting sound sustainability criteria for biofuels, but at the same time should address the issue of fossil fuel subsidies and put a price on CO_2 emissions. The car, aviation and maritime industry, refineries and distribution networks have to evolve together with bioliquids. The current trade barriers obstructing the trade of biofuels and feedstock should be gradually dismantled. Beyond economic factors policymakers should consider ethical and environmental issues as well. The Nuffield Council of Bioethics offers a complex set of ethical and sustainability principles for policymakers:

1. Biofuels development should not be at the expense of people's essential rights (including access to sufficient food and water, health rights, work rights and land entitlements);
2. Biofuels should be environmentally sustainable;
3. Biofuels should contribute to a net reduction of total greenhouse gas emissions and not exacerbate global climate change;
4. Biofuels should develop in accordance with trade principles that are fair and recognise the rights of people to just reward (including labour rights and intellectual property rights);
5. Costs and benefits of biofuels should be distributed in an equitable way;
6. If the first five principles are respected and if biofuels can play a crucial role in mitigating dangerous climate change then, depending on certain key considerations, there is a duty to develop such biofuels.[36]

[36]A. Weale et al., *Ethical framework, Biofuels: Ethical Issues* (London, United Kingdom: Nuffield Council on Bioethics, 2011), Ch. 4.

Fields of Cooperation Between the U.S. and the EU on Biofuels

The U.S. and the EU share a common interest in improving their energy security and reducing their dependency on imported oil. Biofuels play a central role in this plan. Together the two players represented 55 percent of global biofuel production and more than 60 percent of consumption in 2010.[37] They have the highest R&D expenditures in the sector and nearly all new demonstration plants are located in these countries. The U.S. and the EU apply the most sophisticated policy tools and have a vital role in regulating international trade in biofuels. The cooperation between policymakers in the European Union and the U.S. is essential to allow for the sustainable future development of the biofuel industry.

Both the U.S. and the EU have to invest significantly more in the *research, development and demonstration* of advanced biofuels. Support schemes should be feedstock and technology neutral. Government supports have to be predictable and be automatically phased out if a technology reaches maturity. The U.S. and the EU should enhance their cooperation on the R&D of advanced agricultural technologies and practices. As developed nations and technology leaders in the biofuel industry, the U.S. and the EU also share a responsibility for *the promotion of technology and best practices* of feedstock and biofuel production in the world.

The U.S. and the EU should coordinate their action on the implementation of sound *sustainability criteria* for biofuels. The EU Renewable Energy Directive requires biofuels to achieve at least 35 percent greenhouse gas savings versus fossil fuels, rising to 50 percent in 2017 and to 60 percent as of 2018 for new plants. But the mounting evidence suggesting that the European Commission calculated with exaggerated emission-reductions and applied a flawed calculation method, which double counts the carbon savings from certain types of biofuels created an uncertain situation.[38] According to a leaked internal document, the European Commission will probably raise the sus-

[37]Own calculations based on International Energy Agency *World Energy Outlook 2010* and *Technology Roadmap: Biofuels for Transport 2011*.

[38]International Scientists and Economists Statement on Biofuels and Land Use - A letter to the European Commission, October 7, 2011. Available at: http://www.euractiv.com/sites/all/euractiv/files/scientists%20biofuels%20letter.pdf.

tainability criteria to 45-50 percent in 2013 to offset the iLUC effect of biofuels, but will postpone the idea to introduce feedstock-specific GHG reduction targets.[39] The Renewable Fuel Standard in the U.S. requires that at least half of the biofuels production mandated by 2022 should reduce lifecycle emissions by 50 percent, but basically remains silent about the details. The lack of clarity and the uncertainty around sustainability criteria, especially in the EU sends mixed messages to the industry and undermines the credibility of governments' biofuel policies. To avoid such controversies the U.S. and the EU should work together on a *coordinated system of sustainability criteria* that link financial support to GHG performance and incentivize the economical and sustainable use of natural resources.

To facilitate international trade, the transatlantic partners should work together in creating *international certification schemes* for biofuels. These schemes should certify the physical and chemical attributions and the environmental performance of the product, based on life-cycle assessment. Currently there are 67 of such initiatives worldwide, including the European Commission's and the U.S. Environmental Protection Agency's schemes.[40] As the IEA concluded, the "proliferation of standards is increasing the potential for confusion, inefficiencies in the market and abuses such as 'shopping' for standards that meet particular criteria. Such disparities may act as a discouragement for producers to make the necessary investments to meet high standards."[41] The U.S. and the EU should bolster the work of the International Organization for Standardization (ISO), which is currently developing an international sustainability criterion for biofuels.[42] The U.S. and the EU should continue working together on *standardization*.

[39] "EU to delay action on biofuels' indirect impact," *Reuters*, Sep 8, 2011. Available at: http://www.reuters.com/article/2011/09/08/us-eu-biofuels-idUSTRE7874NP20110908.

[40] Dam J. van (2010), Background document from: Dam et al from "The global efforts on certification of bioenergy towards an integrated approach based on sustainable land use planning," in *Journal of Renewable and Sustainable Energy Reviews* (Utrecht, the Netherlands: Utrecht University, 2010).

[41] *Technology Roadmap, Biofuels for Transport* (Paris, France: OECD / International Energy Agency, 2011).

[42] ISO/TC 248, Project committee: Sustainability criteria for bioenergy. Available at: http://www.iso.org/iso/iso_technical_committee?commid=598379.

The Biodiesel Tripartite Task Force and the Bioethanol Tripartite Task Force created by the United States, the European Union and Brazil to align technical specifications for internationally traded biofuels sets a good example.[43]

Standardization is also important in order to step over the "blending wall" and enable the *proliferation of higher ethanol and biodiesel blends*. The 85 percent ethanol blend called E85 is offered at only 2,200 out of 170,000 fueling stations in the U.S.[44] In the EU ethanol is widely accessible only in Sweden, and available in 15 other member states,[45] as well as in Norway and Switzerland.

It is worth considering the *coordination of targets and mandates* applied to biofuels in the U.S. and the EU. Currently the targets of the transatlantic partners are hard to compare and had been established almost exclusively with regard to the development of their respective internal markets.

The U.S. and the EU should *progressively eliminate trade barriers* impeding the international trade of feedstock and biofuels. The U.S. Senate's June 2011 decision ending a 54 cent per gallon tariff on imported ethanol is a positive step towards this direction, which should be followed by the EU as well. Eventually the EU will be particularly interested in the free trade of biofuels, since its domestic biomass-producing capacity is limited.

The progressive *abolition of fossil fuel subsidies* and the establishment of a *global price on carbon dioxide* and other greenhouse gas emissions are essential for the development of all low-carbon technologies, including biofuels. In its New Energy Finance report[46] Bloomberg

---

[43]Tripartite task force Brazil, European Union & United States of America: *White Paper on Internationally compatible biofuel standards*, December 2007. Available at: http://ec.europa.eu/energy/renewables/biofuels/doc/standard/2007_white_paper_icbs.pdf.

[44]Stephanie Dreyer, *Approaching the Blend Wall - What it Means for Our Economic Future*, May 11, 2010. Available at: http://www.renewableenergyworld.com/rea/blog/post/2010/05/approaching-the-blend-wall-what-it-means-for-our-economic-future.

[45]Austria, Belgium, Czech Republic, Finland, France, Germany, Hungary, Ireland, Italy, Latvia, Lithuania, the Netherlands, Poland, Spain, United Kingdom.

[46] "Subsidies for Renewables, Biofuels Dwarfed by Supports for Fossil Fuel," *Bloomberg New Energy Finance*, 29 July 2010. Available at: http://bnef.com/PressReleases/view/123.

Table 4. World biomass shipping

Exporter		Product		Importer
Brazil	→	Ethanol	→	EU, U.S., Japan
Malaysia, Indonesia, Argentina, U.S. (through Canada)	→	Vegetable oils and biodiesel	→	EU
Argentina	→	Vegetable oils and biodiesel	→	U.S.
U.S., Eastern Europe and Russia, Argentina, South Africa, Australia	→	Wood pellets	→	EU
Canada	→	Wood pellets	→	U.S.
U.S.	→	Wood pellets	→	Japan

Source: Based on Biofuels for Transport, Technology Roadmap, OECD / International Energy Agency, 2011.

estimated that global fossil fuel subsidies are twelve times higher than the aggregated government spending on renewable support. In this respect, the leader's statement on the G20's Pittsburgh Summit[47] to phase out fossil subsidies in the medium term was an important step, but unfortunately did not set any concrete deadline.

The current revision of biofuel policies on both sides of the Atlantic justifies a stronger international cooperation. There are a number of already established forums for this dialogue. The transatlantic partners have to make better use of the current framework provided by the *Transatlantic Economic Council*, the *Transatlantic Business Dialogue*, the *Transatlantic Legislator's Dialogue* and the *EU-U.S. Energy Council*. The EU-U.S. Energy Council deals with the sustainability of biofuels on workshop-level, since its foundation in 2009. The last ministerial meeting on November 19, 2010 tasked the body's working group with "exploiting the lessons learned from projects for biorefineries using lignocellulosic and algal feedstocks."[48] Research and development efforts are coordinated through the *EU-U.S. Task Force on Biotechnology Research*, which has a working group dedicated to the transatlantic scientific cooperation on bio-based products, including biofuels.

[47] *Leader's Statement*, The Pittsburgh Summit, September 24–25, 2009. Available at: http://www.g20.org/Documents/pittsburgh_summit_leaders_statement_250909.pdf.

[48] *Joint Statement Following the U.S.- EU Energy Council Ministerial*, Lisbon, Media Note, Office of the Spokesman, Washington, DC, November 19. 2010. Available at: http://www.state.gov/r/pa/prs/ps/2010/11/151185.htm.

Potential Geopolitical Implications of the Increasing Use of Biofuels

Energy security and volatile crude oil prices have always been a major concern for policymakers in the United States and the European Union. Gradually decoupling the increasing energy consumption and economic growth from oil imports would have considerable geopolitical, economic and environmental benefits. But efforts made by the United States and the European Union to reduce the transport sector's dependency on oil entails that "oil producing countries can legitimately stake a claim to increased security of demand through a diversification of customers."[49] It is hard to forecast the economic effect of increased biofuel-consumption in the future. The U.S. Department of Agriculture's study assumes that "imports of crude oil would fall by 16-17 percent in 2022. As a result of lower demand and a decline in the import price, the U.S. import bill for crude oil would decline by \$61-\$68 billion."[50] As a consequence of EU biofuel policies the International Food Policy Institute found that "some countries may experience small negative effects, particularly oil exporters (-0.11% to -0.18% of real income by 2020) and Sub-Saharan Africa (-0.12%) due to the fall in oil prices and rise in food prices, respectively."[51]

Until the proliferation of electric or plug-in hybrid electric vehicles, biofuels provide the only real alternative to reduce the transport sector's dependency on oil. Their high energy density, transportability and compatibility with current distribution infrastructures and vehicles make them an ideal substitute for oil. Another advantage of the biofuel industry lies in the diversity of feedstock, conversion technologies and final products. That flexibility enables biofuel production to be adjusted to local and regional conditions.

---

[49]Claude Mandil, Adnan Shihab-Eldin: *Assessment of Biofuels Potential and Limitations* (commissioned by the International Energy Forum).

[50]Mark Gehlhar, Ashley Winston, Agapi Somwaru: *Effects of Increased Biofuels on the U.S. Economy in 2022* (Washington DC, United States: United States Department of Agriculture, Economic Research Service, Economic Research Report Number 102, 2010). Accessible at: http://www.ers.usda.gov/Publications/ERR102/ERR102.pdf.

[51]Perrihan Al-Riffai, Betina Dimaranan, David Laborde, *Global Trade and Environmental Impact Study of the EU Biofuels Mandate* (Brussels, Belgium: International Food Policy Institute / European Commission, 2010).

"Whether or not biofuels play a significant role in the future energy supply mix depends on the development of biofuel production that avoids or lowers food vs. fuel competition while also contributing to environmental goals."[52] Advanced biofuel technologies offer significant improvement in GHG balance and the development of new energy crops will increasingly enable feedstock production to take place on marginal land. But advanced technologies need considerable support from governments, a favorable regulatory environment, and strong cooperation between major economic players to reach industrial scale and to proliferate around the world.

[52]Govinda R. Timislina & Ashish Shreshta, *Biofuels: Markets, Targets and Impacts* (Washington DC, United States: Environment and Energy, Development Research Group, The World Bank, 2010).

Chapter Nine

After Fukushima: The Future of Nuclear Energy in the United States and Europe[1]

David Koranyi

Before the Fukushima nuclear crisis, the energy world was abuzz with talk of a nuclear renaissance. Since the Fukushima accident, nuclear energy policy is being rethought across the world. The outlook of nuclear energy looks gloomy. The International Energy Agency's World Energy Outlook 2011 introduces a 'Low Nuclear Case', where the total amount of nuclear power capacity falls from 393 GW at the start of 2011 to 339 GW in 2035, compared with an increase to 638 GW in the New Policies Scenario, that is a drop of around 15 percent[2].

The various approaches that countries are taking towards nuclear energy in the wake of Fukushima can be put into five categories. The first category includes those countries with existing nuclear capacity, remaining to be scheduled for rapid expansion, such as China, India, South Korea and Russia. The second category includes emerging countries with new nuclear programs, such as the United Arab Emirates, Saudi Arabia or Vietnam, that are likely to go forward with the projects. The third group includes nuclear energy users that had

---

[1] The study draws heavily on the report of the transatlantic nuclear energy conference co-organized by the Center for Transatlantic Relations (CTR) at the Paul H. Nitze School of Advanced International Studies Johns Hopkins University, EU Center of Excellence Washington, DC and the Atlantic Council of the United States on May 31, 2011. The conference report was authored by David Koranyi and edited by Blythe Lyons, Mihaela Carstei, Wilfrid Kohl and John Lyman. The report is available at http://transatlantic.sais-jhu.edu/.

[2] Reuters: Post-Fukushima nuclear generation could fall 15 percent. http://uk.reuters.com/article/2011/11/04/uk-energy-iea-nuclear-idUKTRE7A 32AV20111104.

Figure 1. World Nuclear Power Reactors

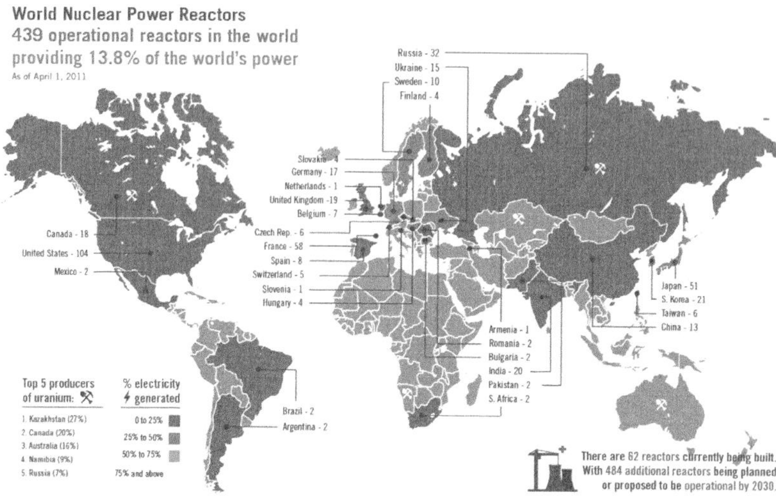

Source: Disaster Preparedness and Visualization, available at http://www.emergency-response-planning.com/news/?Tag=Radiation.

decided to undertake new programs before Fukushima, such as the U.S., the UK or central European countries, but that will likely have a de facto moratorium in place for some time due to political, economic and technical uncertainties related to the accident. The fourth category includes countries that have decided to phase out their nuclear energy program, such as Germany, Spain, Belgium or Switzerland. Countries that have long opposed the use of nuclear energy (like Austria) and now feel vindicated belong to the final category.

First among the decisive factors determining the future of nuclear energy in a transatlantic context is the political dynamic fuelled by changing public perceptions of nuclear energy. An Ipsos-Mori poll taken in the wake of the Fukushima accident shows that public acceptance for nuclear power dipped considerably in most developed countries, although to a differing degree (more in the EU, less so in the U.S.[3]). As more and more details become known about the Fukushima accident, general support could slip further and the "not in my back-

[3] Available at http://www.ipsos-mori.com/researchpublications/researcharchive/2817/Strong-global-opposition-towards-nuclear-power.aspx.

Figure 2. Levelised Costs of Electricity for Different Studies

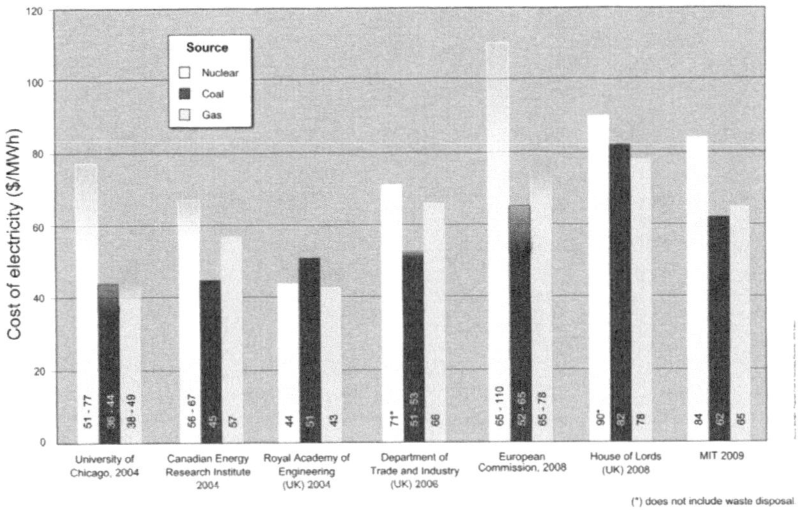

Source: *The Cost of Generating Electricity*, Royal Academy of Engineering, London, United Kingdom 2004; *The Economic Future of Nuclear Power*, University of Chicago, United States 2004; *Levelized Unit Electricity Cost Comparison of Alternate Technologies for Baseload Generation in Ontario*, Canadian Energy Research Institute, Calgary, 2004; *The Energy Challenge*, United Kingdom Department of Trade and Industry, London, 2006; *Energy Sources, Production Costs and Performance of Technologies for Power Generation, Heating and Transport*, European Commission, COM(2008)744, Brussels, 2008; House of the Lords, *The Economics of Renewable Energy, 4th Report of Session 2007-08*, Vol. I: Report, Select Committee on Economic Affairs, London, 2008; *Update on the Cost of Nuclear Power*, Massachusetts Institute of Technology, Cambridge, MA, 2009—graph available at http://en.wikipedia.org/wiki/File:Nuke,_coal,_gas_generating_costs.png.

yard" mentality would likely grow stronger both in the U.S. and Europe: a potentially very serious obstacle to any nuclear development in the future.

Equally important will be the financial implications of enhanced safety standards and other consequences of the Fukushima accident. Affordability matters most when it comes to electricity production. Even before March 2011, there were serious economic obstacles to nuclear development, especially in the U.S. A Massachusetts Institute of Technology (MIT) study[4] showed that nuclear cannot compete with

[4] *Update on the Cost of Nuclear Power* (Cambridge, MA: Massachusetts Institute of Technology, 2009).

coal and gas in competitive, deregulated markets,[5] because the cost of capital for nuclear is somewhat higher than for coal or gas due to the long investment lead time and greater associated risks. The current low costs of natural gas in the U.S. have greatly reduced the nearer term incentive to invest in new nuclear facilities and have reinforced the U.S. public's unawareness about key energy efficiency, sustainability, reliability and security challenges. Furthermore, increased capital costs coupled with the economic crisis and lackluster growth prospects could force governments and private companies to abandon at least some proposed projects.

Finally, uncertainties surrounding global efforts to fight climate change further increase unpredictability. Whereas Europe put a price on carbon, developed an internal carbon market and is at the vanguard of international climate change struggles, the U.S. until now has failed to act at the federal level. Therefore negative externalities such as greenhouse gas emissions are not accounted for in cost calculations, giving an arguably unfair advantage to coal and gas in the U.S. over nuclear (and renewables).

Nuclear Power Policies in the United States and Europe After the Fukushima Accident

United States: Nuclear Remains Part of the Energy Mix

The United States is the world's number one nuclear energy producer. Its 104 operating nuclear power plants accounted for 20% of electricity produced, and for about 10% of the nation's electric generating capacity. Nuclear power accounts for almost 70% of the fuels that do not emit greenhouse gases in the U.S. generation portfolio.[6]

In the past, a number of factors have facilitated nuclear power plant construction in the United States. Consolidation of utilities and standardized plant design, in addition to tax incentives and loan guaran-

[5] The National Academy of Sciences recently found that nuclear is competitive with coal only if there is a requirement for CCS. "America's Energy Future: Technology and Transformation" (2009) available at http://www.nap.edu/catalog.php?record_id=12091.

[6] Energy Information Administration (EIA).

Figure 3. Nuclear Generation 2010—Top 10 Countries: 2,229 Billion Kilowatthours

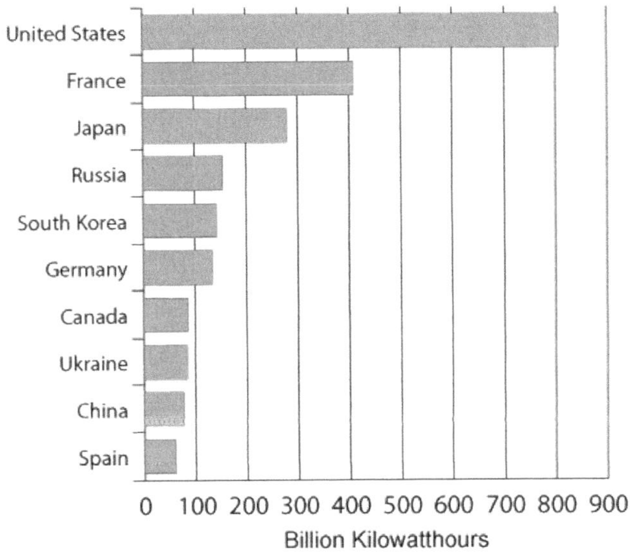

Source: International Atomic Energy Agency, Power Reactor Information System File.

tees in the 2005 Energy Act, have assisted project development. In states that still have regulated electricity markets, nuclear has received support. The Energy Information Administration (EIA) forecasts a small increase in nuclear generation going forward, but by 2030 the share of nuclear in overall electricity generation may actually decrease, outstripped by growth of natural gas fired plants and renewable electricity generation capacity and the continuing dominant but slightly reduced share of coal.

There are seventeen reactor projects for which the Nuclear Regulatory Commission has received Combined Construction and Operating License (COL) applications, but five have been suspended. Currently, there remain twelve potential reactor projects in the United States with active plans for seven units for a total of 8460 MW gross. Active site work is being conducted on four AP 1000 units by the Southern Nuclear Operating Company and South Carolina Electric and Gas. The Tennessee Valley Authority (TVA) has announced plans to complete the Bellefonte reactor project. TVA is completing the Watts Bar Unit 2—the only reactor under construction in the United States at present—and will finance the Bellefonte project with a lease-back sale of Watts Bar Unit 2 and a gas-fired power plant.

New construction is supported by an updated Nuclear Regulatory Commission (NRC) licensing framework, the Energy Policy Act of 2005, and state regulatory frameworks, which have eased the financing process in some states. NRC revamped its licensing process in 1989 for new nuclear reactors to avoid many of the challenges and delays encountered in the past. Although the public can intervene at various stages, the hurdle for intervention increases as the construction process moves forward.

The Energy Policy Act of 2005 provides a number of incentives for both regulated utilities and merchant generation companies that are building new nuclear reactors. The Loan Guarantee Program administered by the Department of Energy, production tax credits, and Federal Standby Support are all incentives that are geared towards helping finance new construction and mitigating investor concerns about investment recovery. Various states with regulated electric markets, mainly in the southeastern region of the U.S., have developed regulatory frameworks to support new nuclear construction. These frameworks include the pre-approval of budget costs and construction schedules, periodic progress review, the collection of financing costs during construction, final prudence determinations, and the ability to abandon investment. However, this only applies to regulated electric utilities, and not to merchant generation companies, which face difficult economic challenges in states with competitive markets.

While nuclear safety has always been a primary concern in the United States, the accident in Japan brought these issues to the center of public attention. Though safety measures continued to improve in the last two decades, in the wake of the Fukushima accident, President Obama asked the NRC to conduct a comprehensive review of the safety requirements of the 104 operating reactors in the U.S., and to strengthen those rules if necessary to safeguard health and safety in emergencies. The industry generally supports this evaluation but notes that there are major operational differences between the U.S.

and Japan, including differences in the preparations for severe accident management and operations, in independent regulation and oversight, in required training, drilling and exercises.

The Obama Administration has made a clean energy economy a top priority and underlined its continuous support for nuclear energy as a clean energy resource. Advancing energy innovation and diversifying the energy portfolio to include more low or zero carbon energy resources are key elements of the Administration's strategy. The Administration, with bipartisan support in Congress, wants to promote investment in next generation clean energy technology including nuclear. U.S. President Barack Obama, in his 2011 State of the Union address, outlined a vision of doubling the amount of electricity generated from clean sources from the current 40% to 80% by 2035. The President has made it clear on several occasions since Fukushima that nuclear energy remains an important part of the energy mix.

European Union: Stress Tests, while Member States Remain Free to Choose Nuclear

Member states (MS) of the European Union freely decide on their energy mix, including the role of nuclear power; they are neither obliged nor prevented from using nuclear energy. At the same time, common rules govern the use of nuclear energy in the framework of the European Atomic Energy Community (Euratom).

The majority of the EU MS support the use of nuclear energy, with fourteen states currently operating one or more nuclear power plants. Nuclear energy accounts for one-third of electricity generation in the EU, and Europe's nuclear industry is a world leader in technology, safety, and security of installations.

Nuclear energy fulfills a key role in the EU's energy mix, and contributes to the security of supply, the competitiveness of the economy, and the fight against climate change. Thus it plays an important part in the EU's climate change and energy policy (the so called 20/20/20 Strategy). Though MS are free to determine their energy mix, according to the 1957 Euratom Treaty, certain aspects of nuclear power fall under collective responsibility. With one of the most advanced legal frameworks in the world, and by setting the highest standards of safety and non-proliferation, EU institutions play an important role in guaranteeing nuclear safety. The EU laid the foundation for these measures at the Union's inception before the disasters of Fukushima or even Chernobyl. Euratom was initially created to coordinate MS research programs for the peaceful use of nuclear energy. The Euratom Treaty today helps to pool knowledge, infrastructure, and funding of nuclear energy. It ensures the security of atomic energy supply within the framework of a centralized monitoring system.

Nuclear energy use in the EU before Fukushima was on an upward trend, but the events at Fukushima inevitably raised some concerns. Discussion within national governments and at the highest levels in Brussels began to focus on the role of nuclear energy in the MS. European leaders decided to launch a major safety review of the 143 reactors in Europe between June 1 and December 31, 2011. The so-called "stress tests" include everything from natural disasters and airplane crashes, to terrorist attacks, and are being conducted in cooperation with the European Commission, the operators and independent national regulators. The three-step process entails a pre-assessment by the operators themselves, with a detailed questionnaire on safety measures and plans (phase I), which will be analyzed and reviewed by the national regulator (phase II) and then peer-reviewed by multinational teams, largely composed of European Commission experts and representatives from other regulatory agencies, and concluding with on-site inspections (phase III). The European Commission will subsequently produce a report with recommendations that will be brought before the European Council.

France—Devoted to Nuclear Energy

France has been at the spearhead of nuclear energy among EU MS, and continues to believe that nuclear is a crucial part of the energy mix. The promotion of nuclear power originates in the proactive response to the general scarcity of resources in the country, and the 1973 oil shock in particular. In response, the government launched an ambitious nuclear program. Today 58 reactors produce more than 75% of France's electricity needs. Political consensus among all governments prevailed on the nuclear program, and was supplemented by other policy initiatives promoting sustainable development, recycling nuclear materials, and nuclear waste management. The nuclear program allowed France to significantly reduce greenhouse gas emissions and served as an environmental asset. In 2005 the National Assembly adopted a law on energy policy orientation, which confirmed nuclear energy's pivotal role in the country's energy mix. The law sets out three key objectives for nuclear energy: ensuring energy independence and security of supply, environmental protection, and economic competitiveness. It also contains the legal framework of the construction of Areva's first European Pressurized Reactor (EPR) in Flamanville.

In the wake of the Fukushima accident, the French government requested the Safety Authority to conduct an audit on the fifty-eight reactors within France, and to focus on the potential risks and operational management issues in emergency situations.[7] In the medium term, France is confident that nuclear energy will play an important role in the global energy mix and aims to promote the highest levels of nuclear security throughout the world. France has used its G-8 and G-20 presidencies this year to further that goal. The June 2011 IAEA conference attempted to draw initial lessons from Fukushima, and to consider steps to enhance the safety regime and collective ability to respond to a serious accident. Nonetheless, France strongly believes that each national authority must remain independent and no supranational authority shall undermine the responsibility of the national safety authorities' prerogatives.

France remains strongly committed to nuclear for the time being and French companies are hopeful that their state-of-the-art nuclear reactor designs could actually benefit from increased safety standards after Fukushima. However, there is an increasing shift in French public opinion over nuclear energy that could be problematic over the medium term.

UK: Replacement of Nuclear Plants and Possible Expansion

The final report on the implications of the Fukushima accident for the UK nuclear industry was published in September 2011.[8] The findings conclude that there is no reason to curtail the operation of UK

---

[7] The Safety Authority will publish its final report by the end of November 2011—after the publication of this book.

[8] Available at http://www.hse.gov.uk/nuclear/fukushima/final-report.pdf.

operating sites, although operators should continue to follow the founding principle of continuous improvement. There are no fundamental weaknesses in the UK nuclear licensing regime or the safety assessment principles that underpin it, and the intention to create the Office for Nuclear Regulation (ONR) in statute will further enhance confidence in the UK's regulatory regime. The final report also confirms that there is no reason to revise the strategic advice given by the regulators on which the Nuclear National Policy Statement was based, or any need to change present siting strategies for new nuclear power stations in the UK. Nevertheless the findings of the report contain a list of recommendations for improvements in emergency response, contingency planning, communications, reviewing flooding studies, reviewing electricity and cooling supplies and new standards on spent fuel strategies.

Nuclear facilities in the UK receive no public subsidies, so any that are pursued are done so by private companies. The government does not know exactly how many plants and reactors will be built, but the intention is that all of the 19 reactors due to be retired by 2023 will be replaced.

Germany: Phasing Out Nuclear Power by 2022

Germany is at the other end of the spectrum. Although German scientists were leaders in nuclear research in the 1930s and contributed to the discovery of nuclear fission, and in the early postwar period the German government gave strong support to research on civilian nuclear power, the accident at Chernobyl had a large negative impact as the fallout reached Germany. This was the beginning of a strong grass roots anti-nuclear movement.

In the 1990s, 23 nuclear reactors generated 40% of electricity in Germany. However, Germany began a change of course with the 1998 SPD-Green coalition government, which agreed to a gradual nuclear phase-out. This was followed by a law passed in 2001, taking effect in 2002, prohibiting construction of new reactors and limiting production of nuclear electricity to 2.6 GWh (a target that would have been reached around 2025).

A new and ambitious energy concept emerged, aimed at replacing nuclear power with renewables over time. In 2001, only 6% of electricity came from renewables (mostly hydropower). With the new concept, increased renewable subsidies, and the so-called "feed-in tariff" for green electricity, the proportion of electricity from renewable sources was slated to increase to 17% by 2010. According to this plan, the ratio will further increase to 35% in 2020 and to 80% in 2050. This would also result in a drastic cut in emissions, of approximately 35% by 2020 and 80% by 2050. Germany is successfully fulfilling its obligations under the Kyoto Treaty, as emissions are already down by 23%.

Based on a new risk assessment after Fukushima, the German government led by Chancellor Angela Merkel decided to speed up the nuclear phase-out process. After Chernobyl, the government believed what happened with the Soviet reactor was highly unlikely to happen in the U.S., Japan or Germany, where reactor designs were more advanced. That risk assessment is different now after Fukushima; it supports the view that no complex, cutting-edge technology is completely immune from complication, and even the remote risks with nuclear installations are just unacceptably high.

Chancellor Merkel's government initially viewed the role of nuclear energy as a bridge technology to a future dominated by renewables. The CDU-FDP government had even decided in 2010 to support an extension of the life of some German reactors, but Fukushima provoked a repeal of that decision and a return to the previous timetable whereby all reactors will be shut down at the end of their life by 2022. In light of Fukushima the seven oldest reactors have been shut down since March 2011, and as of this writing one more is scheduled to shut down promptly. Out of the remaining seventeen, some will close gradually while nine will operate until 2021-22.

At the beginning of 2011, 22.5% of electricity in Germany came from nuclear (about the same as in the U.S.); this will be reduced to zero in 2022. National consumption today oscillates between 40-80 MW a day. German electricity capacity including nuclear is 93 MW, and without nuclear, about 83 MW. Germany plans to increase its share of renewables by investing €1 billion/year in the grid in the coming decade. As of now 7.5% of overall electricity generation comes from wind, 4.5% from biomass, 3.2% hydro, and 1.2% solar.[9]

[9] Eurostat.

In the short and medium term, wind will be the most important source, coming mostly from offshore wind farms on the North Sea. If wind's success remains as it is today, its share of generation capacity could go up to 22%. On the other hand, in the long term, solar energy likely will be the main source, especially given its popularity and falling costs. The German government believes that even if electricity costs increase, German consumers are willing to pay a price to support renewable energy development.

Germany does not intend to increase the use of coal, and in the case that Carbon Capture and Storage (CCS) proves to be a unviable technology, Germany will aim to have coal completely disappear from the mix by 2050. However, this will not be easy, as coal currently accounts for 47% of total electricity generation.[10] Coal use may even increase in the near term, before increased gas supplies from Russia will be available, with the abrupt shut down of a major portion of nuclear generation. Natural gas currently comprises 11% of Germany's electricity generation[11] and is highly likely to increase at least in the medium term. Gas is certainly the best option of all fossil fuels from an emission standpoint but could hinder the development of renewables, just as current technology bottlenecks like energy storage: pumped-storage opportunities are limited in Germany, so they cooperate with Switzerland and Norway. Germany is supportive of fusion power as the ideal solution, but rather skeptical whether a breakthrough could happen anytime soon.

Germany considers the decision to do away with nuclear and increase renewables as a win-win, as it will lead to enhanced energy security, lessen imports on fossil fuels, save about \$30 billion in imports of oil and gas by 2020, and create jobs in the domestic economy. Since 2001, 350,000 jobs[12] were created in the renewables sector, which is the fastest growing sector of the German economy.

It will be interesting to watch the German experiment unfold in the coming decade. Doubts linger over Germany's decision to completely phase out nuclear power by 2022 and over the country's ability to push

---

[10]Eurostat.

[11]Eurostat.

[12]Available at http://www.unendlich-viel-energie.de/de/politik/10-jahre-eeg.html.

renewables forward at such quick pace. Many question the wisdom of the decision since the surrounding states (France and the Czech Republic) might end up exporting more nuclear-generated electricity to Germany. There are worries about the potentially detrimental effects on German competitiveness, as the increased use of renewables could drive up energy prices. Nevertheless, removing nuclear power from the available options inevitably sends a strong signal to the private sector to invest in renewable energy technologies and their deployment.

Nuclear Development Plans Elsewhere Within the EU

In Italy, nuclear power had been used until all the plants were closed down by 1990 in the wake of the Chernobyl disaster and the subsequent referendum on the use of nuclear power. The decision was reversed in 2008 and Italy planned to build 10 new reactors, with the goal of increasing the nuclear share of Italy's electricity supply from today's 10% to about 25% by 2030. After Fukushima, the Italian government put a one-year moratorium on plans to revive nuclear power and through a referendum on June 12, 2011 an overwhelming majority of voters (94%, with 55% of the eligible voters participating) voted in favor of the construction ban.[13]

Switzerland equally reversed course in May 2011 with the Swiss government deciding to abandon plans to build new nuclear reactors. The country's five existing reactors will be allowed to continue operating, but will not be replaced at the end of their life span. The last will go offline in 2034.[14]

Spain has eight reactors producing 20% of the country's electricity. The government decided on a gradual phasing out of all reactors shutting down all nuclear power plants at the end of their lifetime.[15]

Belgium also decided to phase out nuclear power from 2015 but without a final deadline.

[13]OECD Nuclear Energy Agency—available at http://www.oecd-nea.org/general/profiles/.

[14]http://www.nytimes.com/2011/05/26/business/global/26nuclear.html?_r=2.

[15]OECD Nuclear Energy Agency—available at http://www.oecd-nea.org/general/profiles/.

Sweden is heavily dependent on nuclear power—nearly half of the electricity comes from ten nuclear power plants. The country formerly developed a phase-out policy regarding nuclear energy but allowed for the replacement of existing reactors in 2009. The Fukushima accident fundamentally altered prior support of nuclear power (now 64% of Swedes oppose new reactors and only 27% support them). The government so far resisted calls to reverse course again, but facing tough resistance they might be taking a somewhat more careful approach in the future.[16]

Finland approved plans in July 2010 to build a sixth and seventh nuclear facility as a way of reducing carbon intensity and limiting dependency on imported hydrocarbon resources. Finland's energy policy seeks to increase use of nuclear power due to Finland's lack of hydrocarbon resources, little potential for wind, and maximized hydropower potential. On the environmental side, for Finland to meet the EU's 20/20/20 energy and environment targets, the country must increase its renewables to 38% of total energy sources, and reduce its carbon emissions by 20%.

In central and eastern Europe, Poland recently decided to build two nuclear plants with four reactors in total by 2020, which will produce 10% of Poland's total electricity. Coal generates 90% of electricity in Poland, therefore the country is strongly incentivized to develop nuclear to honor binding commitments to the EU to reduce CO_2 emissions. How recent major shale gas findings in Poland could upset these plans remains to be seen. Lithuania had nuclear facilities between the 1980s and the present time, Lithuania's two nuclear reactors provided over 80% of the country's electricity. With the shutdown of the first plant in 2007 and the second in 2009, Lithuania today has no operating nuclear power capacity, and the country depends on Russia for 100% of its imports of gas for electricity generation. Hungary, Slovakia, the Czech Republic and Bulgaria are also at various stages of building replacing and expanding nuclear capacities. The central and eastern European countries are driven by the need to decrease dependency on Russian gas and some are also eyeing increasing export opportunities to the German market.

[16]OECD Nuclear Energy Agency—available at http://www.oecd-nea.org/general/profiles/.

The Future of Nuclear Energy in the U.S. and Europe: Conclusions and Recommendations

We are still too close in time to the Fukushima accident to assess to its full impact on the future of nuclear energy. It is clear that the accident will have wide-ranging implication for the political debates surrounding the peaceful use of nuclear energy, the fight against climate change, business models in the nuclear industry, technology development and safety issues, other energy sectors from gas to renewables and energy prices.

Nuclear power was growing before Fukushima as a source of electricity generation because of three key drivers: to meet rapidly expanding demand for base-load electric power at an affordable price; to increase the share of low and zero carbon energy resources in the energy mix to fight climate change; and to enhance energy security by reducing dependency on imported natural gas and oil and/or diversify energy sources. These drivers are still valid.

However, the constraints on governmental and private sector decisions in the U.S. and Europe triggered by the Fukushima accident may result in protracted political and regulatory processes that further raise costs for nuclear development to unbearable levels. That would effectively kill the development prospects of the nuclear industry in the U.S. and Europe for the foreseeable future, prolong the extended use of fossil fuels as base-load power and prevent the large-scale reduction of greenhouse gas emissions—a catastrophic scenario by all accounts.

To surmount obstacles and constraints of nuclear energy development, proactive policies are needed on both sides of the Atlantic. Nuclear energy needs to be promoted as provider of base-load electricity closely entwined with increasing renewable generation capacities. The United States and Europe can and should play a pivotal role in advancing the efficient and responsible use of nuclear energy in the U.S., Europe and worldwide by creating a level playing field, enhancing global safety standards and further developing the international regulatory and institutional regime.

First, conscious of the climate challenge it is critically important to create a level playing field among all fuel types by internalizing environmental externalities, first and foremost greenhouse gas emissions.

Figure 4. World Energy Demand by 2035

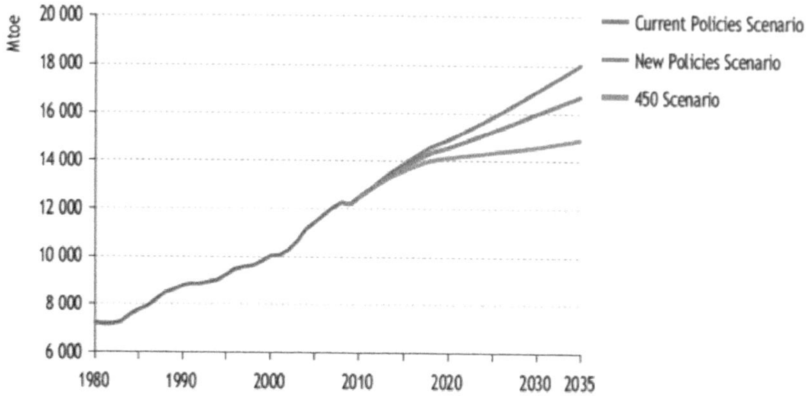

Source: IEA World Energy Outlook, 2010.

The U.S. needs to join the EU in introducing a price on carbon preferably by a cap-and-trade scheme modeled on the EU Emissions Trading System (ETS) and by subscribing to global efforts by signing up to the extension of the Kyoto Protocol and ultimately to a comprehensive and compulsory global effort to fight climate change in the UNFCCC framework. In turn, a charge on carbon would likely provide a major economic incentive to nuclear energy (and renewable) developments.

The next generations of nuclear technologies (GEN III+) will bring fundamental improvements in safety. Nevertheless, as Fukushima proved, emergency preparedness is key and great attention needs to be paid to the eventual loss of onsite power at nuclear plants and the availability of backup power. Prevention is a goal, but rapid response capability must be part of the strategy. It is important to retain and develop the relationships and the depth and breadth of interaction and response that were made during the crisis between all the different stakeholders.

Thus, broader global cooperation in nuclear safety is imperative. The International Atomic Energy Agency (IAEA) must be supported and strengthened in its efforts. The U.S. and the EU should work to strengthen international standards and rules in the framework of the IAEA. Each nation has to have a proper regime for regulation and enforcement of nuclear safety and security. Reviews shall be done at the level where the necessary expertise in the specific reactor designs lays, but it is crucially important to undertake systematic peer assessments.

In the wake of Fukushima, there is a need to rethink how spent fuel is stored at some reactor sites. A reliable emergency plan for coolant loss from spent fuel pools is necessary at every nuclear facility. It is also necessary to rethink the interim storage of spent fuel prior to final reprocessing or permanent storage. It is important to not just improve site-specific plans, but also to establish international and national response capability. Fukushima was an international event. Prevention is a goal, but rapid response capability must be part of the industry. Retaining and developing relationships and the depth and breadth of interaction and response that were made during the crisis between all the different stakeholders are important. Establishing a process and a structure to enable international consultation and collaboration in a time of crisis is crucial. Effective and uniform liability protection is essential: the Convention on Supplementary Compensation for Nuclear Damage (CSC) should enter into force as soon as possible. Such a uniform global legal regime would compensate victims in the event of a nuclear accident. A long-term waste management plan, including guidelines for disposal in a geologic repository should be in place from the beginning of a nuclear power program because reaching agreement on a suitable site for nuclear waste can take a long time.

Turning the tide as regards public opinion is crucial. Government and industry efforts based on independent, reliable and unbiased research and data shall be increased to improve public acceptance of nuclear energy usage. It is critically important to educate the public about externalities, lifecycle costs and overall impact of all energy resource.

Loan guarantees and other federal and state level incentives in the U.S. are critical in helping utilities to obtain financing for new construction and mitigating investor concerns about investment recovery, especially given the size of projects relative to utilities' equity base. Industry efforts to deliver plants on time and on budget are also crucial. A regional approach in central and eastern Europe is needed to avoid costly public investment into nuclear overcapacities and to enhance collective energy security in the region.

The ongoing dialogue in the framework of the EU-U.S. Energy Council should be reinforced that supports the development and deployment of safe, low-cost nuclear technology; aims to develop and promote compatible taxation and incentives policies; and focuses on nuclear waste and site issues. Recommendations based on the findings of the stress tests in Europe and the NRC review in the U.S. will have to be implemented vigorously and without delay.

Participating countries, including the EU, need to continue to properly finance the International Thermonuclear Experimental Reactor (ITER) to prospectively bring fusion energy to the commercial markets.

Last but not least, combating nuclear weapons proliferation remains essential. Any proliferation accident anywhere in the world would have far-reaching consequences also on the peaceful use of nuclear energy. A renewed international framework for peaceful nuclear cooperation is needed to further minimize the risk of proliferation. The idea of an international fuel bank where nations can commercially access fuel needed for the peaceful use of nuclear energy should continue to be pursued vigorously. Discussions are already underway about the elements of such a regime under the International Framework for Nuclear Energy Cooperation (IFNEC). Instead of building new enrichment and reprocessing facilities, fuel leasing services could serve the needs of both the front and back end of the nuclear fuel cycle.

Chapter Ten

Unconventional Gas Resources: A Transatlantic Shale Alliance?

Frank Umbach and Maximilian Kuhn

Although transatlantic cooperation on energy policies has been on the common agenda since the first oil crisis in 1973-74 and led to the creation of the International Energy Agency (IEA) in Paris at that time, the transatlantic partners hardly addressed energy security together throughout the 1980s and 1990s. In the light of liberalized market economies, European governments have mostly seen energy resources only as an economic and not a strategic good. Consequently, the supply of oil, gas and coal resources, and thus energy supply security, has been left largely in the hands of private European companies. Only oil and gas storage was seen as an area where European governments have direct responsibilities. These prevailing assumptions were bolstered by the fact that until recently Europe had not experienced any major energy crisis since the 1970s. As a result, and in contrast to the U.S., where oil supply security in particular has always been a major concern of U.S. government, industry and business circles, European governments did not view energy security as a strategic issue or a challenge.

These circumstances changed with the first Russian-Ukrainian gas crisis in 2006, when EU governments recognized that energy supply security was becoming a major policy challenge, given rising fossil fuel imports and dependencies.[1] A year later, in March 2007, EU member

[1] The following analysis is based on Maximilian Kuhn and Frank Umbach, "Strategic Perspectives of Unconventional Gas: A Game Changer with Implications for the EU's Energy Security." A EUCERS Strategy Paper, Vol. 01, No. 01, May 2011 (London: EUCERS/King's College, 2011). (available at: http://www.eucers.eu/wp-content/uploads/EUCERS_Strategy_Paper_1_Strategic_Perspectives_of_Unconventional_Gas.pdf)—and new information and insights from recent analyses and

states were able to agree on the world's most ambitious "integrated energy and climate policy," which also envisages a proactive energy foreign policy of the EU-27 to promote energy partnerships around the globe with other countries and regions.[2] These energy partnerships also seek to promote the diversification of rising oil and gas supplies from new producer and exporter countries as well as related new transmission supply projects of pipelines, oil and LNG terminals. The *Southern Corridor* project and the *Nabucco* gas pipeline, for instance, are aimed to increase gas imports from the Caspian region as well as from the Middle East in order to reduce the European gas dependency on Russia and its monopoly supplier Gazprom.[3] In this context, transatlantic cooperation on various energy issues, such as oil and gas pipelines, energy efficiency, technology innovations, carbon capture and storage projects, smart grids, energy storage, e-mobility and related climate protection policies have increased since 2006, even though the U.S. and the EU as well as their private energy companies are simultaneously competitors around the globe. Transatlantic cooperation between both sides of the Atlantic has also been institutionalized with the EU-U.S. Energy Council, which has established many common expert working groups and generated transatlantic conferences between official and private actors.[4]

---

conferences, including an expert hearing on the prospects for shale gas in Europe in the European Parliament on September 20, 2011 in Brussels.

[2] See also European Commission, "The EU Energy Policy: Engaging with Partners beyond Our Borders. On Security of Energy Supply and International Cooperation." Communication from the Commission to the European Parliament, the Council, the European Economic and Social Committee of the Regions, SEC(2011)1022/1023 final, COM(2011) 539 final, Brussels, 7 September 2011; and F. Umbach, "Global Energy Security and the Implications for the EU," Energy Policy, Vol. 38, Issue 3, March 2010, pp. 1229-1240.

[3] See also F. Umbach (2011), 'The Black Sea Region and the Great Energy Game in Eurasia', in: Adam Balcer (Ed.), *The Eastern Partnership in the Black Sea Region: Towards a New Synergy*, demosEUROPA, Warsaw 2011, pp. 55-88; and idem (2011), "Energy Security in Eurasia: Clashing Interests," in: Adrian Dellecker/Thomas Gomart (Eds.), *Russian Energy Security and Foreign Policy* (Routledge: Abingdon-New York, 2011), pp. 23-38.

[4] See, for instance, Franklin Kramer/John R. Lyman, *Transatlantic Cooperation on Sustainable Energy Security. A Report of the Global Dialogue between the European Union and the United States*, The Atlantic Council/Centre for Strategic & International Studies (CSIS), Washington D.C., February 2009 and Richard L. Lawson/John R. Lyman/

The global energy future is shaped by unprecedented uncertainties, however, ranging from mitigating climate change to building the right energy mix for each country with the objective of reducing dependencies and gaining energy security. In light of such uncertainties, the emerging global gas market can be summarized in one word: volatile.

As a cleaner-burning alternative to coal and oil for electric power generation, natural gas is an attractive "transition fuel" towards a low-emission global energy mix, a potential which may or may not be fully realized, depending on the evolution of the structure of the gas market itself, as well as on energy policies by key players. Today's distinct regional gas markets—where demand is more or less fully satisfied by national or regional supply—will become more integrated under the impact of the present "gas glut," with more flexible forms of trade, such as liquefied natural gas (LNG), and through the continuing liberalization and integration process of the EU energy markets. Traditional views on geographic distribution and "energy security" are increasingly being challenged, and will require continuous adjustments from industry, governments and new technology providers. Players will need to understand more intimately rapidly changing markets and pricing structures and will have to invest in diversification for both supply and demand to achieve greater strategic flexibility. Changes—irrespective of the outcome—will have a significant impact on transatlantic relations, national economies and, ultimately, on consumers.

In this complex setup it makes particular sense to analyze unconventional gas more closely, not only as the one issue of most interest, but also because it is relatively new to the concept of transatlantic cooperation. Both continents—the U.S. and Europe—have had disconnected and independent natural gas markets in the past, with their own history and industry structure, in which natural gas has been historically transported by pipeline and only since the late 1990s transports from producing regions by ship through liquefaction (LNG) became more common. With the growing role of LNG in the natural gas markets, the U.S. and European markets became more interconnected, especially in their efforts to secure resources in producing

---

Mihaela Carstei, *A Shared Vision for Energy and Climate Change. Establishing a Common Transatlantic Agenda*. A Joint Project of the Atlantic Council of the United States and Clingendael International Energy Program, in cooperation with the Institute for the 21st Century Energy, Washington D.C., March 2010.

countries. Overshadowed by growing LNG markets, further pushing globalization, however, a national development had been ignored by most of the major IOCs until around 2006-2007: the development of unconventional gas in the U.S., in particular shale gas.

Unconventional gas (shale gas, tight gas and coal-bed methane) development is not a revolution but rather an evolution of utilizing modern techniques and combining two key technologies—horizontal drilling and "slick water" hydraulic fracturing—which eventually cracks shale rock and thus cracked the code for opening up major North American shale gas resources. However, the release of unconventional gas resources triggered what can rightly be called a revolution in global gas markets. Unconventional gas not only transformed the U.S. energy market, and in particular the natural gas market, but it was also the tipping point of a fundamental change in global gas markets. An increase in incremental U.S. non-conventional shale gas production coincided with other critical economic, political, and technological factors—the drop of demand linked to the global recession and the arrival of new LNG delivery capacity—which all together created a sudden global "gas glut" and, thus, laid the groundwork for an expanded role of natural gas in the world economy,[5] or a "Golden Age of Gas," as the International Energy Agency (IEA) put it.[6]

In early June 2010, the Bureau of Intelligence and Research in the U.S. Department of State and the U.S. National Intelligence Council invited European energy experts and representatives of European gas companies to Vienna to introduce the American shale gas revolution, its new drilling technologies, and to debate its impact on the European gas market. Since that time, transatlantic discussions and cooperation on shale gas in Europe have increased, and U.S. energy companies increasingly invest in potential European unconventional gas fields. But unlike the North American case, the development of natural gas in Europe and Eurasia depends on the development of a com-

[5] For a detailed report comparing U.S. and European unconventional gas development, including its historic development, see Kuhn and Frank Umbach, op. cit.

[6] IEA, "Are We Entering a Golden Age of Gas?". Special Report. *World Energy Outlook 2011*, Paris 2011.

plex set of cross-border relationships for the build-up of pipelines at the intergovernmental, regulatory and corporate levels.[7]

The following analysis examines the rapidly changing natural gas markets and the role of unconventional gas as well as its impacts on European energy security and transatlantic energy policies. It will highlight in particular the geo-economic and geopolitical implications and discuss whether a "Transatlantic Shale Alliance" is possible.

Why Look at Global Natural Gas Markets?

For decades to come natural gas will be a critical element in the fundamental climate, economic and political calculations made by all major economies, primarily as a transition fuel for a renewable energy mix. Natural gas is and will remain an essential component of the energy mix for a variety of reasons: massively increased global and regional established gas reserves; competitive economic costs; a relatively favorable carbon footprint when properly developed and transported; and a natural synergy with the large-scale development of intermittent electricity sources. Most recently, gas has provided a partial substitute for the loss or delay of (new) nuclear power generation after the Fukushima disaster.

The growth of commercial gas reserves by almost 30% over the last decade is due to the fact that oil companies have begun to search, explore, and produce gas, as well as to technological advances in developing and transporting natural gas. Moreover, natural gas will become more relevant for the renewable energy industry because "Green Gas," or SNG (Synthetic or Substitute Natural Gas), can both provide storage and transport energy by using already existing infrastructure. Natural gas will provide a balancing option for renewable energy and the possibility to store and save electricity through conversion into natural gas. Thus, gas will not only be a bridge to a sustainable future

---

7 See also David Victor, Amy M. Jaffe and Mark H. Hayes (eds.), *Natural Gas and Geopolitics. From 1970 to 2040* (Cambridge-New York: Cambridge University Press, 2006) and Paul Stevens, *The 'Shale Gas Revolution': Hype and Reality*. A Chatham House Report, London, 2010.

energy mix, but also remain a systematic component in the provision of energy security.[8]

Natural gas is currently the world's third largest source of primary energy and gas reserves are more geographically dispersed than oil, with many of the world's top consumers holding significant domestic reserves—especially when taking into account unconventional gas like shale gas, coal bed methane (CBM), tight gas and others. The fact that 63% of gas reserves is located in regions other than the Middle East makes gas more attractive to governments wishing to reduce their energy (i.e. oil) dependency on this presumably unstable region. According to the 2010 BP Statistical Review, proven global (conventional) natural gas reserves in 2010 amounted to 187.1 trillion cubic meters (tcm) with a reserve-to-production (R/P) ratio of 59 years, compared with 46 years (including Canadian oil sands, based on 2010 production levels) for oil.[9] Natural gas may become the fastest growing fossil fuel until 2035, increasing by up to 2% annually.[10]

Natural gas is increasingly seen as a viable source of energy due to improvements in technology, changed regulatory frameworks, and the fact that gas is seen as the "greenest" fossil fuel.[11] Until a few years ago, however, declining indigenous production in the U.S. and Europe, and consumer markets increasingly seeking more distant supplies, meant that LNG deliveries from often remote sources, had became more prevalent. Future dependencies on a few countries—notably Russia,

---

[8] For further reading on the future potential of natural see Robert A. Hefner III, *The Age of Energy Gases: The Importance of Natural Gas in Energy Policy*, speech and paper by Robert A. Hefner III at the Aspen Institute's Aspen Strategy Group's conference "The Global Politics of Energy," Aspen, Colorado, August 2007. Available at: http://www.ghkco.com/downloads/ASG-ImportanceofNaturalGasinEnergyPolicy 08.07.doc

[9] See BP, op. cit., p. 7 and 20.

[10] See IEA, op. cit.

[11] However, even though a new study by the *National Center for Atmospheric Research* has concluded that a greater reliance on natural gas will emit far less CO_2 than the use of coal (up to 50% higher), it would fail to slow down climate change significantly. While coal use and CO_2 emissions are detrimental to the environment, release of comparatively large amounts of sulfates and other particles from natural gas cools the planet by blocking incoming sunlight. Furthermore, analyses of climate change are also complicated by the uncertainty regarding the amount of methane (often seen as much more dangerous than CO_2 for climate change) that leaks from natural gas operations—see "Switching from Coal To Natural Gas Would Do Little for Global Climate, Study Indicates," *European Energy Review*, 12 September 2011.

Qatar, Iran, Turkmenistan, and Australia—that control the bulk of conventional gas reserves would have become more dominant. Energy security issues arose out of concerns over energy imports and threatened independent decision-making mainly of European countries, thus posing a threat to transatlantic strategies. The use of the energy weapon as a political tool created tensions between supplier and consuming countries. But unconventional gas is not only abundant and available all over the world, it also challenges the market power of producer countries as well as potential supplier cartels (such as the Gas Exporting Countries Forum/GECF), and thus strengthens the position of consuming countries.

A New Gas World Arising—
Unconventional Gas as a Major Driver

The advantage of unconventional gas is that it is a domestic source of fuel supply that enhances the energy security of the respective country. As pointed out before, the U.S. unconventional gas success story has been a paradigm shift that has turned expectations upside down. It has essentially been a "game changer" for the emerging world gas market. Developing unconventional gas reserves attracts foreign direct investment (FDI), creates new jobs, and helps to diversify away from other imported fuels, or, in the case of the U.S., help the country gain energy independence.

The reward of accessing these very large unconventional gas volumes is that their potential is vast—it is a resource several times greater in magnitude than that of conventional sources. Estimates of recoverable resources are growing at a greater pace as technological advances permit access to gas from "unconventional" resources. The most prolific shale reservoirs are relatively flat, thick, and predictable; the formations are so large that, once drilled, the wells are expected to produce gas at a steady rate for decades. Generally, it is assumed that shale gas wells flow rates are considerably lower than their conventional peers, but once production stabilizes, a well could produce consistently for 30 years or more.[12]

[12]Joseph H. Frantz and V. Jochen, *When Your Gas Reservoir Is Unconventional, So Is Our Solution—Shale Gas*, Schlumberger, October 2005.

While recoverable conventional gas resources are estimated to amount to 404 tcm, unconventional gas resources are estimated at over 900 tcm (according to the US Geological Survey (USGS) and the German Federal Institute for Geosciences and Natural Resources (BGR).[13] From these 900 tcm, at least 380 tcm appear to be recoverable, which brings the total recoverable conventional and unconventional gas resources to nearly 800 tcm—equivalent to about 250 years of current production.[14] In addition to the U.S., the biggest potential of unconventional gas is currently seen in the region of the former Soviet Union (CIS), Central Asia and China. But given the lack at present of sufficient geological information and credible exploration drilling test data outside of the U.S., prospects for unconventional gas production are likely to remain uncertain for at least the next 2-5 years.

Nevertheless, exploration drilling for shale gas and coal-bed methane has already begun in China, India, Canada, Australia (i.e. coal bed methane production) and Europe (tight gas has been identified in Poland, Hungary and Germany).[15] The U.S. Energy Information Administration (EIA), in its *International Energy Outlook 2010*, estimated that the unconventional gas production of Canada and China will amount to 63% and 56%, respectively, of their total domestic gas production in 2035 (Reference Scenario).[16] The Paris-based IEA, being rather conservative regarding estimates for future worldwide unconventional gas production, expects that around 35% of the global increase in gas production—from 3,149 bcm in 2008 to 4,535 bcm in 2035 (44% in the timeframe)—will come from unconventional gas sources.[17]

---

[13]United States Geological Survey, "World Petroleum Assessment," Boulder, Colorado, 2000; BGR, "Reserves, Resources and Availability of Energy Resources" (Hannover: *German Federal Institute for Geosciences and Natural Resources*, 2009).

[14]IEA (2010), *World Energy Outlook 2010*, Paris (OECD/IEA). See also Alex Forbes, "The Great Potential of Unconventional," *European Energy Review*, December 9, 2009.

[15]IEA, Ibid; BGR, op. cit.

[16]EIA, *International Energy Outlook 2010* (Washington D.C.: EIA, 2010).

[17]IEA, *World Energy Outlook 2010*. An even more conservative outlook is presented by the World Energy Council, "Survey of Energy Resources: Focus on Shale Gas" (London: WEC, 2010).

Figure 1. Regional Distribution of Tight and Shale Gas Resources

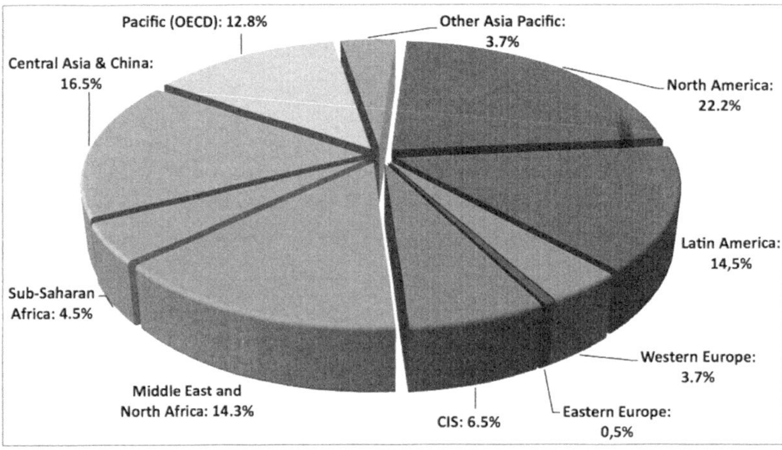

Source: BGR, Reserves, Resources and Availability of Energy Resources. Hannover/Germany 2009, p. 93.

In April 2011 EIA published a commissioned report offering a new assessment of worldwide shale gas resources. The report analyzed 48 shale gas basins in 32 countries, containing almost 70 shale gas formations. However, it excluded other potential regions such as Russia, the Middle East, Southeast Asia, and Central Africa because they either have large conventional gas reserves (i.e. Russia and the Middle East) or lack sufficient information to carry out an initial assessment. Although the report represents "a moderately conservative 'risked' resource" assessment for basins, the findings of the initial assessment conclude that the worldwide shale gas resource estimate adds another 40% to the world's total of technically recoverable gas resources, representing an increase from 16,000 to 22,600 trillion cubic feet (tcf).

The EIA report also concluded that surprisingly China holds technically recoverable assets of around 50% more than the U.S. Although some important regions have not been included, the report's evaluation shows that the assessed worldwide shale gas resources are already

[18]See EIA, *World Shale Gas Resources: An Initial Assessment of 14 Regions Outside the United States* (Washington D.C.: U.S. Department of Energy, April 2011).

significantly larger[18] than assumed in the only previous study conducted by H-H. Rogner in 1997.[19]

The IEA expects that total gas production In China will rise from 80 bcm in 2008, to 140 bcm in 2020 and 180 bcm in 2035 and that the "bulk of the increase" in tight gas, coal bed methane and shale gas is expected within this timeframe. In November 2009 China signed a cooperation agreement with the United States on shale gas development projects. China's National Energy Administration (NEA) is currently drafting a national shale gas development plan that aims for commercial production as early as possible in order to (1) increase cleaner energy consumption and (2) reduce reliance on carbon-intensive coal. Shell is cooperating with PetroChina on this and is currently drilling 17 wells, including some for tight gas and shale gas; BP is also seeking to cooperate with Sinopec on joint shale gas development projects in China. The government has set up special research projects in Beijing focusing on shale gas exploration and development technologies, and plans to invest $1 billion a year over the next five years into shale gas development if the exploration test drilling underway proves to be successful.[20] China's Ministry of Land and Resources has estimated its total unconventional gas reserves at 15-30 tcm, in contrast to the more optimistic EIA estimate of 36 tcm. In June 2011 it held its first auction for domestic gas producers to bid on exploration rights for four shale gas blocks. While domestic gas consumption already surged to 110 bcm in 2010, the NEA plans to double the natural gas share in its energy mix from presently 4% to 8% by the end of 2015.[21]

China is also paying special attention to coal-bed methane (CBM) due to the lower capital requirements, the technological entry barriers in comparison to tight or shale gas exploration and production, and the involvement of many more players. But, while CBM production capacity was only 2.5 bcm in 2009, production volumes are even lower at 0.7 bcm. At present, production targets for CBM were 5 bcm by the end of

[19]H.-H.Rogner, "An Assessment of World Hydrocarbon Resources," *Annual Review of Energy and the Environment* 22, 1997, pp. 217-262.

[20] "China Energy Authority Drafting Shale Gas Development Plan: NDRC," Reuters, March 28, 2011.

[21]See Victor Wang, *Shale Gas to Play a More Prominent Role in China's Energy Mix*, Interfaxenergy.com, September 14, 2011.

2010, and are 30 bcm by 2020 and 50 bcm by 2050. Present production costs are about 50% higher than those of conventional natural gas.[22] Global resources of CBM alone amount to 135.5 tcm—372.5 tcm.[23] While India has also estimated 1.8 tcm of recoverable shale gas reserves and conducted its first exploratory drilling of shale in north-east India last January and plans an additional three in March 2012, experts see unconventional gas as a long-term option rather than an immediate energy source. Like many other countries interested in exploiting their unconventional gas resources, India currently still lacks a regulatory framework for unconventional gas exploration.[24]

Global Foreign Policy Implications of Unconventional Gas

The shale gas buzz is not only about its the radical change it is causing to the energy industry, but also about the its political and international ripple effects. Unconventional gas has become the new 'elephant in the room,' with global geopolitical implications that have caused a chain reaction: European gas prices are being renegotiated and revised. It has also caused an average of 15% of Gazprom's supplies to be delinked from oil-indexation in 2010. But the implications are greater still: relatively cheap and abundant gas, along with the carbon advantage of gas, makes nuclear power and coal relatively more expensive than currently assumed. Indeed, making gas a major transition fuel through 2030 will help renewable energy efforts to reduce emissions, at a lower cost in order to mitigate the impact of climate change.[25]

---

[22] "China Gas Sector. Key Takeaways from the Asia-Pacific Unconventional Gas Summit," Yuanta-Industry Update, April 1, 2010.

[23] BGR (2009), op. cit., p. 95.

[24] See Sara Stefanini, "India's Shale Gas Potential Raises Excitement, but Gas Is Still a Long Way Away,", September 13, 2011, and idem, "India Needs Regulations for Successful Unconventional Gas E&P," *Interfaxenergy.com*, September 14, 2011.

[25] Dieter Helm, *The Coming of Shale Gas: the Implications for Oil and Energy*, http://www.terrafirma.com/Alternative-perspective-page/articles/295.html. This is in contrast to a report from the Tyndall Centre arguing against shale gas in particular as a transition fuel and highlighting the potential risks to human health and the environment. See Wood. R., Gilbert P., et al., *Shale gas: a provisional assessment of climate change and environmental impacts*. A report commissioned by the Cooperative and undertaken by researchers at the Tyndall Centre, University of Manchester, 2011. See also Robert W. Hogath, Renee Santoro, Anthony Ingraffea, *Methane and the*

Unconventional gas has helped to shift the balance from a seller-dominated market to one dominated by buyers, in contrast to the global oil market. Unconventional gas is nowadays the "new policy" option for European countries, giving buyers more leverage to renegotiate high Russian oil-indexed gas price demands that are included in long-term contracts. Thus, unconventional gas, even without being produced in Europe, puts a certain price cap on high Russian gas prices, as it can become a potential source of diversification, particularly if Russian gas prices are higher than the break-even point for European unconventional gas. All this has the potential to make unconventional gas development economically feasible and, politically speaking, more appealing. Unconventional gas, and shale gas in particular, has already become a negotiating tool for Europe in a changing gas market that is enhancing the region's energy supply security by diversifying energy sources and enabling the prioritization of a domestically-located resource. Furthermore, the carbon footprint of domestically-produced shale gas in Europe is estimated to be 30% lower than Russia's long-distance pipeline gas.[26]

The increasing self-reliance of the U.S. energy market—due to unconventional gas and natural gas liquids (NGL) production—has another important economic effect on global competiveness. The widening gap between the North American and the European oil and gas market widens is accentuating competitive differences between the two continents. This could be seen in the recent unrest in the Middle East and North Africa (MENA), which affected the European oil and gas market to a greater extent than it did the U.S. As the U.S. becomes almost self-reliant through unconventional gas production, it becomes

---

Greenhouse-Gas Footprint of Natural Gas from Shale Formations, Climate Change (Springerlink.com), April 12, 2011; critical comments on this "biased" study by other environmental experts and new research projects—see 'Five Things to Know About the Cornell Shale Study', *European Energy Review*, April 27, 2011 (originally in: *Energy in Depth*); Gregory C. Staple and Joel N. Swisher, "The Climate Impact of Natural Gas and Coal-Fired Electricity: A Review of Fuel Chain Emissions Based on Updated EPA National Inventory Data," *American Clean Skies Foundation* (www.cleanskies.org), April 19, 2011; and Mohang Jiang et. al., "Life Cycle Greenhouse Gas Emissions of Marcellus Shale Gas," *Environmental Research Letters*, 6 (July-September 2011).

[26]See House of Commons, *Shale Gas* (London: Energy and Climate Change Committee, 2011), ch. 6, point 154.

increasingly unaffected by world market volatility.[27] As the price dynamics are decoupling, pushed by unconventional gas developments, they are pulled by the developments on the LNG market. With possible future exports of unconventional gas by LNG to Europe or Asia, the hitherto set gas prices would become more connected, transparent and eventually lead to a truly connected global gas market, similar to the oil market, from which all gas consumers will benefit. A global gas market is in line with U.S. foreign policy and its transatlantic allies' interest.[28]

Unconventional Gas Development as a Foreign Policy Tool—the U.S. Global Shale Gas Initiative (GSGI)

With the huge success story of the unconventional gas industry in the U.S., the U.S. Department of State launched the Global Shale Gas Initiative (GSGI) in April 2010 in order to promote its technology and to help countries seeking to utilize their unconventional natural gas resources identify and develop them safely and economically. The GSGI helps countries assess their shale gas potential and provide regulatory guidance on its development. Under the GSGI the U.S. has established partnerships with China, India, Poland, Ukraine, Jordan and other countries.

Hence, energy, and in particular unconventional gas, is becoming a foreign policy tool for the U.S. administration. The U.S. is attempting to achieve various goals at the same time: to promote American technology and gain market shares in other countries; to ally with strategic partnering countries and help them reduce import dependencies; and to push natural gas as a more environmentally friendly fuel, thus supporting U.S. and global efforts to address climate change.

Energy security has been the "Achilles heel" of many countries by forcing them to cooperate with unstable states or rough-regimes and thus limiting the foreign policy leverage of the international community vis-à-vis these mainly rent-driven countries. Unconventional gas

[27] See Ruud Weiermars, "Transatlantic Energy Prices Show Need for Realignment," *Oil & Gas Journal* 2011.

[28] See also John Deutch, "The Good News about Gas: The Natural Gas Revolution and Its Consequences," *Foreign Affairs* 90, 2011, p. 82.

changes this picture. By making abundant unconventional gas resources (which can be found close to many demand centers) available, it reduces such dependencies through diversification, and in addition creates and supports national and local industries. With such a powerful tool at hand, it reduces the need of the U.S. to consult with countries on their energy import strategy from the Middle East and Russia. Instead, the U.S. can promote its technology to help these countries gain more energy independence and help replace coal with natural gas as a cleaner fuel and reduce emissions from coal use. As David Goldwyn, the then State Department's coordinator for international energy affairs, told the *National Journal*: "For energy-security reasons, we're interested in having countries like China and India and Eurasian countries develop their own means so they don't get it from countries of concern for the U.S."[29]

The new U.S. administration's foreign policy of promoting unconventional gas poses a threat for Russia, the world's largest natural gas resource holder. The current gas market situation—marked by a global "gas glut," a de-linkage of gas prices from oil prices, and European pipeline prices being temporarily three times that of LNG spot market prices—already poses a severe problem for Gazprom and, more importantly, for its long-term gas importers. Unconventional gas, furthermore, has the potential to remove Gazprom's near-monopoly of European gas supplies. In the fourth quarter of 2010, for example, Russia's gas exports to Europe declined by 17% owing to market oversupply due to redirected LNG cargoes and unseasonably warm weather. Unconventional gas development could reduce Russia's market share in Europe from its recent peak of 26% in 2007 down to just 13% by 2040.[30]

Therefore, Gazprom needs to diversify as its European export model suffers. It is expected that Gazprom will operate in three distinct markets: (1) the traditional European market; (2) a de-regulated and mixed domestic market; and (3) a new Asian market.[31] However,

---

[29]*National Journal*, November 18, 2010.

[30]See Kenneth B. Medlock III, Amy Myers Jaffe, Peter R. Hartley (2011), *Shale Gas and U.S. National Security* (James A. Baker III Institute for Public Policy, Rice University), p. 45 f.

[31]Kushnir and Kapustina, *Natural (Gas) Partners—One Step at a time for Russian Energy to China* (Frankfurt/M.: Deutsche Bank Research, 2010).

indications for a new eastern strategy for gas supplies to China—a new, big, growing market—might not solve the problem Gazprom might be facing. China is already moving towards a more gas-dependent economy due to several reasons previously mentioned associated with gas as a clean and relatively cheap fuel. But Petrochina's estimates of 45 tcm of Chinese unconventional gas are more than Russia's proven conventional reserves. China also seems more likely to dictate low prices connected to coal or hub pricing than to pay a high premium for Russian gas as the Europeans do.

Consequently, with the high cost of building new infrastructure in China and developing expensive new upstream projects in East Siberia and the Russian Far East, diversification of gas deliveries to China will not allow Gazprom to reduce its exposure to Europe. China's energy security policies have been driven less by short-term profit or value-aggregation objectives than by long-term principles of a sustainable energy security and diversification. Thus, the U.S.-China Shale Gas Resource Initiative—an initiative dedicated to enabling the U.S., as "a leader in shale gas technology and developing shale gas resources,"[32] to help China pursue its diversification objectives for its energy mix and imports in the future. In short, China is far more likely to support its local economy by producing domestic unconventional gas than increase its dependence on very expensive Russian pipeline gas.[33]

The GSGI also affects Iran, the second largest natural gas resource holder, since the U.S. strategy is to encourage other countries, such as China, not to breaking sanctions against Iran and instead to still their hunger for new energy resources by producing their own local unconventional gas resources. It seems that the U.S. strategy is working, as China is delaying its projects in Iran, slowly withdrawing from the Iranian energy sector, and vigorously pushing for unconventional gas exploration and production domestically.[34]

---

[32]The White House, "Statement on U.S.-China Shale Gas Resource Initiative", Washington D.C., 2009.

[33]For a detailed report on China´s unconventional gas exploration and prospects see: Aizhu Chen, "RPT-SPECIAL REPORT: China set to unearth shale power," April 20, 2011, *Reuters* (available at: http://r.reuters.com/jub29r).

[34]E. Downs and S. Maloney, "Getting China to Sanction Iran: The Chinese-Iranian Oil Connection," *Foreign Affairs* 90, No. 2, March 2011, pp. 15-21; and Medlock III, Jaffe and Hartley, op. cit.

Developing unconventional gas resources worldwide increases overall energy security for importing countries with unconventional resources and reduces competition over Middle Eastern resources and over LNG. As a local, decentralized source of fuel, unconventional gas is evolving as a factor balancing potential supplier monopolies and alleviating geopolitical tensions.

Prospects, Challenges and Constraints for Europe

> Given the early stages of unconventional gas development, contract rigidities in European gas markets and an insufficiently integrated transportation network, it is too early to determine the magnitude and pricing impact of unconventional gas on individual EU Member States. What is certain is that approaches to unconventional gas will vary widely between Member States, which are setting their own priorities on energy developments.[35]

While initial assessments of Europe's unconventional gas potential were rather skeptical and very conservative, thereby also focusing on different circumstances for specific countries,[36] Europe also has depositories of significant unconventional gas resources, with estimated total recoverable reserves between 33 to 38 tcm, according to new U.S. estimates.[37] An IHS CERA study of 2010 even concluded that the geological potential in Europe for shale gas, and to a lesser extent for coal-bed methane, may even compare to North America's

---

[35]R.Lawson, W.Ramsey, John R. Lyman, Mihaela Carstei, *European Unconventional Gas Developments. Environmental Issues and Regulatory Challenges in the EU and the US* (Washington D.C.: Atlantic Council, June 2011), p. 4.

[36]Paul Stevens, *The 'Shale Gas Revolution': Hype and Reality*. Rik Komduur (2010), "Europe Not Ready for Unconventional Gas, Yet," *European Energy Review*, June 21, 2010; and Roderick Kefferpütz, "Shale Fever: Replicating the US Gas Revolution in the EU?," *CEPS Policy Brief*, No. 210, June 2010. See also Karel Beckmann, "Shale Gas Will Not Be as Important in Europe as in the U.S." (Interview with Jean-Francois Cirelli, President of Eurogas and President of the French company GDF-Suez)', *European Energy Review*, May 9, 2011.

[37]EIA, *World Shale Gas Resources: An Initial Assessment of 14 Regions Outside the United States*, 2011.

potential and, thus, be "far greater than is commonly understood."[38] Although an unconventional gas boom in Europe is not guaranteed at present, the IHS CERA study expects a reasonable minimum level of 60 bcm of an annual production, whereas the also plausible high scenario could reach even 200 bcm shortly after 2025.[39]

Concessions for shale gas test drilling have already been granted, with the Netherlands, France, Germany, UK, Sweden, Hungary, Switzerland, Ukraine, and Poland at the forefront. But it is presently unclear how much shale the rock formation contains and at what cost the gas can be produced. Meanwhile, fears of contaminating drinking water and other environmentally-related concerns have led to declarations of a moratorium for drilling unconventional gas resources in France, the Netherlands and Germany,[40] whereas Poland, Great Britain, Bulgaria and Ukraine are going ahead with test drillings of their own unconventional gas resources.[41] These declared moratoriums and their justifications are based not only on understandable environmental concerns, but also on myths, misinformation and vested interests and lobbyism of other energy industries (renewables, nuclear, conventional gas etc.):

- most environmental concerns in the U.S. arise from a lack of environmental stewardship from small independents combined with ineffective state regulatory framework and monitoring; EU regulation is more robust and European shale gas

---

[38]IHS CERA, *Rising to the Challenge: Turning North America's Unconventional Gas Supply Potential into Reality*, 2010.

[39]Ibid.

[40]See also Yves de Saint Jacob, "France's 'Green Vote' Kills Shale Gas— and Targets Nuclear Power as Well," *European Energy Review*, July 21, 2011; Peggy Hollinger, "France to Ban Fracking of Fossil Fuels," *Financial Times*, May 11, 2011; David Jolly, "French Lean Toward Ban of a Controversial Gas Extraction Technique," *New York Times*, May 10, 2011 and Stefan Nicola, "German Shale Gas Faces Uphill Battle," *European Energy Review*, April 11, 2011.

[41]See also Katarzyna Kacperczyk, "Shale Gas Development: Polish Perspective," Wilton Park Conference, June 16, 2010; Jan Cienski, "Poland Hopes to Tap Big Reserves of Shale Gas," *Financial Times*, August 9, 2011, Ekke Overbeek, "Shale Gas Doesn't Make Poland the New Norway Yet," *European Energy Review*, June 14, 2011; and Atanas Georgiev, "Shale Gas Battle in Bulgaria—High Stakes for Europe," *European Energy Review*, September 1, 2011.

is more likely to be developed by international oil companies, which have a much better track record in managing environmental impacts, albeit this will come at a certain cost. A European regulatory framework is under way, though it may take some time and might be outpaced by national regulatory frameworks.

- Given the fact that the gas-bearing and water-bearing layers are widely separated below the surface (with the gas being much deeper), hydraulic fracturing of shale gas is unlikely to contaminate drinking water as long as continued care and independent regulation is guaranteed. Up to now there is not one documented case of groundwater contamination due to hydraulic fracture stimulation itself.
- Over time, new well and reservoir management technologies are making it possible to significantly reduce the number of well pads required. Furthermore, those new technologies will widen the production base of unconventional gas all the time, while "weakening the strategic importance of conventional reserves and the power of those who hold them."[42]
- As U.S. experiences indicate, deep shale gas uses four times less water than coal and up to two thousand times less than biofuels.[43]

In the forthcoming years, unconventional gas could also become a very important factor for Ukraine's efforts of diversifying its gas imports in order to reduce its high dependencies on Russia. In November 2010, the Ukrainian Ministry of Environment and Natural Resources and the National Joint Stock Company (NAK) "Nadra of Ukraine" declared to have the biggest, or one of the biggest, shale gas deposits in Europe. The Ukrainian government seeks to investigate the potential volume of shale gas by mid-2012, assumed to be between 10-30 trillion cubic meters (twice as large as those of natural gas), and has invited international investors to analyze and develop the Ukrain-

[42] "Coming Soon To a Terminal Near You," *The Economist*, August 6, 2011.

[43] See Wolf Regener, "BNK's Shale Gas Projects in Europe," BNK Petroleum, Wilson Park Presentation, June 2010, p. 15.

ian shale gas deposits.[44] In February 2011, at the Strategic Partnership Commission meeting of the U.S.-Ukraine Energy Security Working Group, both sides signed a 'Memorandum of Understanding' to establish a framework for technical cooperation that will assess the unconventional gas resource potential in Ukraine. This agreement includes the involvement of the U.S. Geological Survey (USGS), which is currently undertaking a global unconventional gas resource assessment.[45] Meanwhile, Total, Eurogas (a U.S. company), and Royal Dutch Shell have announced to conduct test explorations and feasibility assessments.

Against this background, together with the fear in Moscow of losing further markets shares in its most important export market for conventional Russian gas, as well as the geopolitical games resulting from Gazprom acting as the spearhead of Russian foreign policy, it is hardly surprising that representatives of the Russian government and Gazprom try to downplay the importance of shale gas in Europe and point to negative implications of unconventional gas production in Europe for its environment and the EU's climate mitigation efforts.[46]

Summary and Perspectives for a Transatlantic Shale Alliance

Given worldwide and European prospects for unconventional gas production, it is evident that the availability of even a fraction of unconventional gas potential for European and other energy markets makes it unrealistic to argue that the EU, at present, needs all the gas pipelines (i.e. Nabucco and Whitestream) currently being discussed or new LNG-terminals. Both the European gas industry and the EU member states need to prioritize the most economical and energy

[44]"Ukraine Claims to Possess World's Biggest Shale Gas Deposits," *PR Newswire*, November 29, 2010.

[45]"U.S.-Ukraine Unconventional Gas Resource MOU Signed," *Embassy of the United States*, Kyiv, Ukraine, February 15, 2011.

[46]"Alexander Medvedev Answers Your Questions—Part One," *Financial Times*, February 18, 2011; "Gazprom Chief Steps Up Attacks on Shale Gas," *Financial Times*, February 18, 2011, "Gazprom Chief Calls Shale Gas a 'Bubble," *Financial Times*, February 18, 2011, and Andrey Konoplyanik, "The Economic Implications for Europe of the Shale Gas Revolution," *Europe's World*, January 13, 2011.

security enhancing pipelines, while at the same time following the same rationale when considering the options for the new re-gasification terminals that would facilitate higher and more flexible LNG imports during times of crisis. In this regard, unconventional gas as a domestic source may definitely further enhance the EU's future energy supply security, although the prospects for significant unconventional gas production presently do not appear to be a concrete option before 2020.

The U.S. unconventional gas success story represents a paradigm shift that has turned expectations upside down. In essence, it has been a game changer for the emerging world gas market. The historic development of both the U.S. and the European gas market has shown that the markets are becoming interconnected and globalized through LNG, and that unconventional gas provides the local content to the equation. But the U.S. experience also underlines the fact that the U.S. shale gas revolution, or at least its huge impact, is only possible with a liberalized gas market and "would not have occurred in North America if access to market was blocked by pipeline transportation monopolies."[47]

The U.S. shale gas boom enabled a revolutionary domino effect on the European market, with the contractual structure, based upon 25-year long term take-or-pay oil, and linked natural gas contracts that had hitherto dominated the market being re-negotiated. Consequently, shale gas is already having an increasing influence on European gas prices and is expected to continue doing so through 2015. By taking a more strategic perspective, unconventional gas in Europe also serves the three major objectives of the "energy triad" or "energy trilemma":

- Energy Supply Security: As a domestically-produced energy resource, unconventional gas is increasing national and EU-wide energy security by diversifying the national and EU energy mix as well as by decreasing gas imports from unstable or politically problematic suppliers. But even in the case that no unconventional gas will be produced in Europe itself, its expected increased production outside of Europe will offer

---

[47]Kenneth B. Medlock III, "Impact of Shale Gas Development on Global Gas Markets," *Natural Gas & Electricity*, April 2011, pp. 22-28(23).

many more LNG options for future EU gas imports and, thus, also contribute to Europe's future energy security.

- Economic Competitiveness: The present situation implies that unit supply costs will likely be higher in Europe than in the U.S., but also much lower than Russia's future long-distance pipeline gas from new and very expensive gas fields in the high north (such as Yamal) or even the Barents Sea and the Arctic, based on long-term contracts with inflexible price adaptation mechanisms and highly problematic third-party clauses. Furthermore, with continuous technology innovations of hydro-fracturing technologies in the years ahead, production costs in Europe will also go down as predicted by all historical experience with new energy resources and drilling technologies.
- Environmental/Climate Protection: Given the conclusions of new environmental studies and in contrast to populist environmental myths and ill-informed public opinion, shale gas—like conventional gas—produces equally lower CO_2 emissions than coal. Moreover, the carbon footprint of domestically produced unconventional gas is also approximately 30% lower than long-distance Russian pipeline gas if one includes full life-cycle emissions. Negating domestically produced shale gas means higher imports of pipeline gas and LNG, which neither increase energy supply security nor contribute much to climate mitigation efforts. In addition, domestically produced unconventional gas is both technologically and environmentally less risky than the increased drilling of conventional gas resources in ever deeper seas or even in the environmentally most sensitive Arctic and Antarctic regions.

Nonetheless, unconventional gas exploitation is still at an embryonic stage and needs further development. But European governments should not focus just on myths and misinformation as well as short-sighted considerations, but rather take a longer-term strategic perspective by assessing and recognizing the full potential of the global geoeconomic and geopolitical impacts on their own energy security, such as the many benefits for their economic competitiveness.

Regardless of how the concrete outlook for European unconventional gas development looks—and whether or not unconventional gas will become affordable and sustainable in the mid-to-long term in Europe—shale gas has already changed the European market, even before a single well has been drilled or a single molecule of unconventional gas extracted from European basins.

In light of this situation, a reaffirmation of a transatlantic energy alliance would symbolize the continuation of a long history of transatlantic energy cooperation, which includes the nuclear energy cooperation in the 1960s and 1970s and the close ties regarding the hydrogen-based economy in the 1980s. EU-U.S. energy cooperation has always been mutually beneficial and has reminded both sides of their common interests. A new transatlantic shale gas alliance would reconnect this tradition and could contribute to a whole set of policy issues: new sources of foreign direct investment, creation of new jobs and infrastructure, diversification of gas imports in Europe, and in the case of the U.S., help to become more energy self-sufficient.

But U.S. and European energy policies also need to recognize that they can also clash with each other on unconventional gas because European countries, leading in renewable industries (such as Germany), are promoting these industries not just for their own internal market, but also increasingly abroad as part of their future export and industrial strategies. A transatlantic shale gas dialogue will inevitably flesh out the contrasting evolution between both continents. Energy dependency has shaped and influenced foreign, economic and domestic policy on both sides of the Atlantic. However, there has been one major difference in the EU energy policy approach compared to the U.S. approach: the EU has focused more on the collective and demand-side, whereas the U.S. has rather focused on the supply-side.

Given the dual challenge to global energy security and climate change, however, both sides need both to promote renewables and to develop unconventional gas. Thus, a transatlantic strategy should focus on a combined promotion strategy to cope with the dual challenges ahead. Policymakers from the EU and the U.S. will need to reconcile their domestic political agendas with their economic interests for their future gas market policies and their energy security strategies.

Chapter Eleven

What are the Security Challenges of the Natural Gas Golden Age? Natural Gas Security in the U.S. and Europe

Kornél Andzsans-Balogh[1]

Natural gas is a depletable, fossil, natural resource. Compared to oil it has lower energy per unit content, however it has relatively low carbon footprint—lowest among fossil fuels—making it an attractive energy resource, which could technically replace oil in many modes of consumption. Nevertheless, the transportation and storage of natural gas remains more difficult and expensive then oil, casting a shadow over the huge potential of natural gas. In some crucial areas natural gas has overtaken the position of oil in the last three decades such as in the heating and electricity generation sector of the Organization for Economic Cooperation and Development (OECD) member countries. Both in the United States and Europe the majority of new electricity power plants built since 1990s were natural gas fueled, so-called Combined Cycle Gas Turbine (CCGT). No new nuclear power plant was built in the last two decades in these regions. Irrespective of whether transportation sector switches to electrcity or natural gas as a fuel, in both cases this gaseous primary energy source will play a crucial role in the future. Natural gas clearly pushes to take a leading role. As electricity network balancing depends progressively on natural gas generation capacities—partly due to increasing share of hardly controllable or storable renewable energy production—and more households depend on natural gas run heating, the short interruptions in natural gas supplies are increasingly inconvenient and have escalating economic, political, and social side effects. This chapter therefore discusses the

[1] The contents of this paper are the authors' sole responsibility. They do not necessarily represent the views of the Regional Centre for Energy Policy Research or any of its members.

medium term supply security risks of natural gas.[2] Long run security is not discussed due to the substitutability of natural gas, which limits long term policy decision making. To put it other way, any interruption within a year in supplies are desirable to be avoided, but if it is clear that natural gas supply will not be available from a supplier in a foreseeable period, both policy makers and the market is capable to adapt to that new situation by providing alternative fuel sources.

How is Domestic Natural Gas Production Reshaping Supply Security in the Transatlantic Space?

Today, the United States is the largest consumer and producer of natural gas resources. Meanwhile, Europe is the second largest natural gas consumer of the world. In 2009 the EU imported 65% of its natural gas demand; more than 70% of imports were shipped via pipelines. The largest pipeline import partners were Russia (34%), Norway (31%) and Algeria (14%).[3]

Historically, the United States and the European Union have been the largest natural gas consumers in the world.[4] Both entities have significant yet depleting conventional domestic natural gas production. The 1990s saw a significant upward demand trend for natural gas, as electricity generation from natural gas became more efficient, and commercial, industrial consumers also started to switch to this "convenient" fuel. The opening of new domestic natural gas production wells were not capable to keep up with the pace of the increasing demand. Many wells peaked in this decade in the U.S., in the Netherlands and the North Sea. Due to the fact that domestic production was not satisfactory on the long run and its replacement doable with imports, natural gas started to gain a geopolitical importance both sides of the Atlantic.[5]

---

[2] Short term risk measurement has technical tools. For a summary of the existing measurement tools and methodology see Regional Centre for Energy Policy Research: *Measures and Indicators of Regional Electricity and Gas Supply Security in Central and South-East Europe* in: Péter Kaderják ed., *Security of Energy Supply in Central and South-East Europe* (Budapest: AULA, 2011), pp. 8-50.

[3] Eurostat (2009).

[4] British Petrol: *BP Statistical Review of World Energy 2011* (London: BP, 2011).

[5] The European dependence increase on Soviet natural gas supplies were already heavily debated in the US during the 1970s and 1980s.

At the millennium the majority of market participants expected that the United States would substitute declining domestic production with natural gas imports from overseas.[6] The global energy majors rushed to support leading LNG import projects to cover future U.S. demand.[7] However, thanks to the favorable regulatory environment in the U.S., the solution for an adequate supply answer came from small-scale unconventional natural gas producers that were overlooked by the big energy majors for decades. The unconventional gas production technology—fracturing—was there since the 1950s. However, only upward trends in U.S. domestic natural gas prices at the end of the 1990s and the experience gained in the last decades made this technology feasible. The fracturing and cracking technology debuted as a major "game changer" in the American continent. In 2009 the U.S. was capable of covering 87% of its demand domestically.[8] Today and in the upcoming two decades unconventional natural gas production is the basis of the security of natural gas supplies in the U.S.[9]

On the other side of the Atlantic Ocean Europe's domestic production is less fortunate. The traditional conventional fields in the Netherlands and the North Sea are depleting. There have been insights that the unconventional natural gas production revolution can happen in the European Union as well. According to available data, the EU's unconventional natural gas technical potential is 1.4 trillion cubic meters.[10] The largest fields of these types are in Poland, France, Denmark, the UK and the Netherlands. Some have argued that the

---

[6] International Energy Agency: *World Energy Outlook 2000* (Paris: IEA, 2000), Energy Information Administration: *U.S. Natural Gas Markets: Recent Trends and Prospects for the Future* (Washington: U.S Department of Energy, 2001).

[7] David G. Victor, Amy M. Jaffe, Mark H. Hayes, eds., *Natural Gas and Geopolitics From 1970 to 2040* (Cambridge: Cambridge University Press, 2006), Energy Information Administration, *The Major's Shift to Natural Gas* (Washington: U.S. Department of Energy, 2001).

[8] U.S. Energy Information Administration, *Annual Energy Outlook 2011* (Washington: EIA, 2011).

[9] Kenneth B. Medlock, Amy Myers Jaffe, Peter R. Hartley, *Shale Gas and U.S. National Security* (Washington: James A. Baker III Institute for Public Policy and U.S. Department of Energy, 2011), available at: http://bakerinstitute.org/publications/EF-pub-DOEShaleGas-07192011.pdf.

[10] U.S. Energy Information Administration: *World Shale Gas Resources: An Initial Assessment of 14 Regions Outside the United States* (Washington: EIA, 2011).

complexities of European regulation and legal systems are not as favorable as that of the United States, i.e. that ownership rights, tight environmental regulation and lack of experience make Europe a less favorable place for shale gas investments.[11]

Does Europe Really Not Welcome Shale Gas Investments?

Property rights are a major obstacle to shale gas investments in Europe. While in the U.S. the land owner owns everything under ground, in Europe mineral resources under the ground belong to the state. As land owners cannot gain automatic benefits, they are less motivated to allow major drilling developments on their field. This leads to a "not in my backyard" syndrome. While investors have to pay rents for the government, they also have to agree on land usage with the owner. Europe, particularly the former EU15,[12] have a very strong tradition in environmental protection and politicians are keen to be careful not to take a stand on issues that cause public skepticism.[13]

Unconventional natural gas has been perceived as a water- and chemical-intensive industry that threatens drinking water resources and requires huge areas of land. These prejudices have to be overcome on the continent. Furthermore, there is no consensus whether the availability of waste amounts of cheap domestic natural gas should be perceived as a threat to renewable energy production development or rather perceived as a catalyst that helps to regulate the imbalanced energy systems that renewable energy production causes. While large scale, widely adaptable electricity[14] storage technology is not readily

[11]Florence Gény: *Can Unconventional Gas be a Game Changer in European Gas Markets?* (Oxford: Oxford Institute for Energy Studies—Working Papers NG46, 2010) and Paul Stevens: *The 'Shale Gas Revolution': Hype and Reality* (London: Chatham House Report, 2010).

[12]Austria, Belgium, Denmark, Finland, France, Germany, Greece, Ireland, Italy, Luxembourg, Netherlands, Portugal, Spain, Sweden, United Kingdom.

[13]See Fukushima consequences in the European electricity market.

[14]Although hydro energy provides storage opportunity its usage and capacity is limited to geographic givens. The European hydro potential has been developed, further development potential is limited to the Balkan countries or natural parks and protected areas. For example in Hungary a debate is going on for several decades whether the country neeads a pumped stroge hydroenergy. See more: Andras Kiss et

available, some renewable energy production remains weather-dependent (wind flows, sunny hours), thus the amount of their production cannot be regulated. The clear advantage is that natural gas production can be activated fast, and the fuel itself (natural gas) can be stored, while large scale non-hydro dependent electricity storage remains unresolved. Thus natural gas could serve as a bridge technology until the renewable energy or electricity storage becomes a mature technology.

The environmental worries have a particular importance for the new eastern European members that are basically the Achilles Heel for EU level natural gas supply security. Eastern European members are more relaxed on environmental issues, and they are heavily dependent on one single natural gas supplier. Security of supply, due to lack of diversified supplies are top of the agenda that is far from the case on the western side of the European continent. Furthermore, for eastern European countries unconventional gas production would be a significant boost for the economy and would provide faster integration with the old EU member states.

Nevertheless, environmental concerns within western European member states can lead to such EU level regulations that may not favor unconventional production in the whole EU. France has already suspended rights for unconventional gas test drillings for an unknown time. Germany has not expressed a clear stance yet, while Poland is particularly against introducing any EU-level legislation that would shrink its maneuvering opportunity or constrain an investment-friendly environment for shale gas development. Major pipeline importers such as Russia would clearly benefit from a strict environmental regulatory outcome.

Europe also lags behind the U.S. in unconventional drilling technology and human know-how. Europe lacks the production capacity of drilling wells and wellheads to support an unconventional production boom. Historically, the continent has no onshore drilling expertise that matches in scale that of the U.S. The lack of human capital and industrial capacities means that even if the regulatory environment is welcoming the necessary capacity build-out would take years. This means that unconventional natural gas production would not

al, *A szivattyús energiatározás kérdésének közgazdasági elemzése* (Budapest, REKK, 2008) Available at: http://www.rekk.eu/images/stories/letoltheto/wp2008-8.pdf.

provide a full covered domestic supply security earlier then 2020, but more likely around 2030.[15]

Furthermore, the huge unconventional natural gas technical potential does not reflect the economically viable potential of Europe. Economic viability is much dependent on pricing and technology costs. Currently there are no major signs that would suggest booming natural gas prices that would initiate unconventional gas production. Hungary can stand for an example of peculiar shale gas expectations in Europe. In 2008 this central European country became a place to hype unconventional natural gas, when ExxonMobil and Hungarian MOL announced test drilling of the Mako field, which has a potential of 350 bcm of shale gas. The test drilling results were a disappointment; the field with its deepness, pressure and temperature characteristics was very challenging, the flow rates turned out to be very low. After two years the companies decided to abandon the project until new technology became available. Today the Mako field concessions are partly held by the Serbian NIS, in which Gazprom has a majority stake. Thus, the Hungarian example shows that technical potential may turn out to be uneconomic and that traditional pipeline suppliers are keen to secure their comfortable supply positions by acquiring stakes in potential upstream fields that can lead to competition.

In short, it can be said that the shrinking European domestic production cannot be replaced by unconventional resources currently and it is questionable for the present whether after 2020 unconventional resources could play a major role in European supplies. The domestic and EU-level regulatory regime and environmental concerns are significant challenges for unconventional natural gas production. The capacity buildup in Europe will be drawn out. Furthermore, these weaknesses will be actively or indirectly exploited by the largest natural gas suppliers in the European neighborhood. The conclusion is that European demand on the mid run can be satisfied only by expanding imports. Therefore, the security of LNG and pipeline

[15]In Poland commercial domestic unconventional gas production is expected to begin by 2014 (in Hungary small scale uncovnetional production is already online). However, the capacity build out will take at least a decade. AFP: *Poland targets 2014 for shale gas debut* (Warsaw, AFP, 2011).

imports will play an increasing role in European policy decisions in upcoming decades.

How Secure are LNG Supplies?

Natural gas liquefaction technology, similarly to unconventional natural gas production, has a long history, yet the evolution of this technology and its spread happened at a snail's pace. The technology has been available since the 1940s. However, it was hugely expensive, and due to its technical complexities (e.g.: natural gas has to be cooled below -160 Celsius) and limit of physics, it remains a capital intensive technology and no significant cost reduction is expected.[16]

Large scale LNG shipments started from Algeria in the 1970s, targeting the U.S. and the European markets. However, since then many producers built out import capacities. Currently Qatar has the largest liquefaction capacity and it has approximately accounts for one fourth of total capacity in the world. Nevertheless, the announced new capacity development suggests that Qatar will lose its number one position and that Australia will become the largest liquefied natural gas exporter on capacity basis by the end of this decade.[17] This will allow a slight balancing out for the supply diversification and minimize the role of a single huge supplier that can influence pricing and may lead to sensitive geopolitical areas.

Despite Qatar's current significant influence it can be said that LNG suppliers are well diversified. Qatar supplies both the Atlantic and Pacific basin markets, these markets have a spread of suppliers from South America, the Caribbean, Africa to the Far-East. This not only guarantee global access for LNG, but also limits the single one large suppliers role. In this manner, most potential turbulence in any one LNG producer country could be managed in the upcoming decade. The global economic slowdown and the U.S. shale gas revolution enabled the industry to keep up with the pace of demand growth.

---

[16]International Energy Agency: *Energy Technology Perspectives 2010* (Paris: IEA, 2010).

[17]Howard Rogers, *LNG Trade-flows in the Atlantic Basin: Trends and Discontinuities* (Oxford: Oxford Institute for Energy Studies—Working Papers NG41, 2010).

At the moment and in the midterm there are no foreseenable bottlenecks on the supply side.

LNG is supplied by ships, thus securing the major shipping routes in Hormuz and in Asia (South China Sea, Paracel and Spratly Islands) will remain a key issue. These shipping routes cannot be secured by the European allies alone without the active support of the United States. Threats to Asian supplies will automatically lead to rising prices in European markets. In Asia, China's LNG demand is projected to grow enormously. It is likely that China and Japan will each seek to have their supply routes secured, and Australia as a supplier will take an increasing role in the upcoming decade. Nevertheless, for Europe, the Middle Eastern and African shipping routes remain crucial to guarantee physical delivery of LNG shipments.

Besides securing physical deliverability of LNG, another important aspect is avoiding extreme price volatility. Although LNG is perceived as a form of global trade of natural gas, its product pricing remain dependent on the shipping distance. Thus shipping from the Pacific Basin producer to the Atlantic Basin consumer in most cases is not economic. In the last few years, however, when LNG markets—due to economic slowdown and North American unconventional production boom—have seen turbulent times of oversupply, some LNG cargoes even from Australia landed on European shores.[18] Despite this extreme example, in the long run these physical shipments are hard to justify, due to the increased long distance shipping costs. In general, swap deals should happen between the Pacific and Atlantic Basin producers to limit the arbitrage gap. Thus the "global LNG pricing" is secured up until both basins have sufficient surplus capacities. Any major interruption or capacity drop out in one of the basins can lead to significant price signals in the other market. It can be said that European import pipeline infrastructure, which many Asian countries lack completely, is not sufficient to fully hedge against Asian price developments. Hence, while Asia is more vulnerable to LNG supply security, the EU as a growing natural gas importer shall not downplay the effects of turmoil, increasing demand and security challenges in the Far Eastern LNG market.

[18]Eurostat and LNG trader reports.

In the Atlantic Basin the shale gas revolution in the U.S. caused significant shifts in the LNG market. By becoming self-reliant on domestic production, the United States no longer needed those LNG import capacities that were specifically built out for U.S. market demand during the last decade. As a consequence, natural gas from these surplus capacities started to flow to the nearest and biggest market, Europe. Some argue that these surplus capacities will be swallowed up by the Asian markets during the next decade.[19] Currently we saw that the events in Fukushima did not affect the midterm supplies to Europe. After a few months of significantly increased prices—when the increase of oil prices also played a role—the European spot prices returned almost to their pre-Fukushima level. It can be expected that Australia's heavy investments will meet most of the future Pacific Basin LNG demand. Thus, the demand increase in Asia will absorb the European surplus slower than earlier expected.

Secondly, increased liquidity in the Atlantic Basin provided the opportunity for functioning moments of the two Atlantic markets as theoretically one single market, i.e. arbitrage became possible between the two markets through LNG swaps. In the past decade, the European and the U.S. spot market prices became linked twice for a longer period.[20] In a very simplified way this means that the price differential between the U.S. and European markets is equal to the transportation and storage cost. Otherwise the upward and downward terms and scale are equal. In this framework the U.S. price is set by the demand and overwhelmingly by the cost of domestic production, while the European price is more complex with Take or Pay pipeline supplies, domestic production and LNG imports. In theory, when the U.S. producer has an opportunity to enter the European market with profit (e.g.: LNG swap deal), that is production cost and transport is lower than the European spot price, it enters the European market capping the European price increase level up until the limit of available surplus

[19]See more on this discussion, Ibid.

[20]Directorate-General for Energy: *Quarterly Report on European Gas Markets* (Brussels: DG-Energy, 2010) pp.4., Jonathan Stern and Howard Rogers: *The Transition to Hub-Based Gas Pricing in Continental Europe* (Oxford: Oxford Institute for Energy Studies—Working Papers NG49, 2011) and Howard Rogers: *LNG Trade-flows in the Atlantic Basin: Trends and Discontinuities* (Oxford: Oxford Institute for Energy Studies—Working Papers NG41, 2010).

capacity in the U.S. Or the other way around, when Europe has extra surplus from cheap pipeline natural gas below the U.S. price plus transportation cost, the gas flows to the U.S. market.[21] This way the price of the two markets follow each other, provided there are no bottlenecks at the supply and receiving terminals. Once the U.S. builds out its first LNG export capacities in the Atlantic Basin this theoretical framework will be further strengthened.

While it appears that there is a liquid natural gas supply to Europe, it is limited to the western European natural gas markets. The EU has to ease access to this liquidity by eastern European countries—in some cases land-locked—that do not have access to LNG terminals.[22] Brussels has already set plans for upgrading and increasing internal pipeline capacities. Maintaining the current liquidity level and eliminating physical and regulatory bottlenecks will limit oil-price-linked pricing and increase market liquidity in the new member states.[23]

---

[21]See a more sophisticated description in Rogers 2010, Ibid.

[22]There are several plans in Eastern Europe to build new LNG terminals:

- Romania and Bulgaria both plan to have an LNG or compressed natural gas (CNG) terminal on the Black Sea cost. However, no energy major is involved in these projects besides the national champions. Taking account the shape of these countries budget and the rising capital costs these plans cannot be seen as real.
- In Croatia the Krk LNG terminal has been in the news for two decades. Although international majors, like E.On are present in the project company that is part of the EU backed North-Sout corridor of the Visegrad4 countries (Czech Republic, Hungary, Poland and Slovakia) the lack of demand, budget problems in the region and involved energy majors, the competition from South corridor pipelines (see last part of this chapter) all compose a significant barrier for investment decision. Therefore it is higly unlikely that an LNG terminal would become available before the end of this decade.
- The only LNG terminal under construction in Eastern Europe is located in Swinoujscie, Poland. Although, post 2014 this 2.5 bcm receiving capacity would increase Polish supply security, it will not guarantee it, as its capacty is minor compared with the country's consumption of 16.4 bcm in 2009. With such capacity the Swinoujscie LNG terminals regional contribution is limited. It has to be mentioned that capcity increase up to 7.5 bcm have been announced, but this security contribution have to be judge on a border perspective. Namely, what path will Poland choose to address its climate commitments and whether it will switch its overwhelmingly coal based energy prodcution to a more environmental friendly one. (data obtained from www.polskielng.pl).

[23]Bulgaria, Czech Republic, Estonia, Hungary, Latvia, Lithuania, Poland, Romania, Slovakia, Slovenia.

These commitments will bring the first results in 2015. In general, it is expected that the price link between the two shores of the Atlantic Basin markets will grow. That leads to an increasing likeliness that the U.S. and European markets become one common natural market, which would mean certain obligations for policymakers in the field of diplomacy and common transatlantic policy decisions.

The common market, which brings benefits for both regions' end users in a form of low and competitive prices, will exist up until the adequate liquidity levels are guaranteed. An export restriction on U.S. shale gas, would lead to economic inefficiencies. U.S. producers would receive less value for their product, and domestic consumption—which does not reflect global prices—would lead to wasteful utilization. Although it can be argued that domestic production is a crucial element of national security, not having a transatlantic common natural gas market with America's closest allies could lead of endangerment of their supply security. Thus an America-first approach may carry a higher geopolitical price tag at the end.

The United States and the European Union are taking a leading role in energy market liberalization. The U.S. and the UK set a benchmark of national energy market opening, which is why EU efforts lead to an example of how national economies can create a single energy market, thus laying the example of a more opened global energy market. The U.S. shale gas revolution boosted western European natural gas market liquidity and thus significantly accelerated market opening efforts in Europe. As a result, natural gas markets in the transatlantic space are showing the signs of becoming a single one. This integration is driven by market forces, with active support of policymakers. The market integration and the increased liquidity give solutions for such long-standing geopolitical challenges as the Russian energy supply dominance or Maghreb energy supply dependence in Europe. These indirect (market-driven) political efforts provide greater diplomatic maneuvering room by relaxing some painful interdependencies. However, they also underscore the need for a more sophisticated mix of soft and hard policy tools to maintain the fragile and newborn transatlantic natural gas market.

The transatlantic single natural gas market cannot become a reality without LNG supply; however, LNG is not sufficient to meet all

European demand. While LNG is the key to enabling market functioning and liquidity in the transatlantic space, pipeline supplies will remain a crucial form of supply of the Eastern side of the future common transatlantic market.

How to Deal with Pipeline Security?

Natural gas supplied via pipeline creates a significant interdependence between the producer and supplier. The pipeline shipment, unlike LNG, cannot be switched between different producers and different markets. The involved parties engage each other—and sometimes the third parties as transit countries—in a long term commitment. David G. Victor et al. have looked at the history of natural gas deliveries for the last four decades.[24] Their research concluded that in case of pipeline delivery, producers have higher propensity to cancel deliveries and to underutilize pipeline than the propensity of end users cancelling deliveries due to price or delivery condition dispute. In addition to these academic findings, the European experience with regard to pipeline security is closely associated with the 2006 and 2009 January events, when Russia decided to stop shipments for Ukraine after several unsuccessful rounds of negotiations with their Ukrainian counterparts regarding future delivery amounts and prices. That led to the largest natural gas supply crises in European history. The episodes significantly damaged Russia's reputation as a reliable supplier, which previously had been immaculate, at least among the 15 old EU member states.[25]

The share of natural gas pipeline imports in the United States is minimal. Any major interruption in the North American import network can be addressed with sources from domestic storage, production or additional LNG imports. From Washington's view point the key question remains how secure is the pipeline supply of its Euro-

---

[24]David G. Victor, et. al, op. cit.

[25]The Swedish Defense Research Agency documents 38 cases of supply cuts. Jakob Hedenskog and Robert Larsson. *Russian leverage on the CIS and the Baltic States*. (Stockholm: User reports. FOI - Swedish Defence Research Agency, 2007) Available at: http://www2.foi.se/rapp/foir2280.pdf.

pean allies, as well how much strength and influence producer countries may gain by supplying the European markets.

At least 70% of European natural gas imports came via pipeline in 2009. The three key suppliers were Algeria, Norway and Russia; their supply did not overlap with each other in some parts of the continent, thus making some EU members dependent on one single pipeline supplier. Algerian gas is delivered mainly to the Mediterranean markets of Spain, Portugal, France and Italy. In these regions Algerian gas has to compete with other LNG or North Sea suppliers. Furthermore the LNG-receiving terminals in this region are around 50-80% utilized,[26] offering a wide enough buffer to receive any additional LNG supply in case of pipeline disturbances. An interruption of Algerian pipeline supplies due to internal EU network characteristics would be felt only in a few pocket regions of the EU, and network development projects are addressing these solvable problems.

Norwegian gas also enters the more liquid, competitive parts of the European market. Although any major disturbance or loss of 31% of supply would be challenging to substitute from other suppliers, both Norway and Algeria have additional LNG terminals for exports. In this manner the mode of delivery can be diversified up to existing capacity limits. In contrast, Russia is enjoying a more comfortable supplier position, having its own exclusive markets in eastern Europe, where it only has to compete with local domestic production. Furthermore, these eastern European markets have less weight, both measured in the amount of consumption and in political power. This gives Russia significant space to maneuver in terms of price setting and bargaining conditions. In northern and western European markets Russia is one out of many competitors and has no bargaining leverage; it has to follow market trends. In Germany, Russia's unwillingness to introduce more flexible terms of pricing caused headache, financial and market share loss to German energy majors (RWE, E.ON), however, the domestic prices could not be influenced by the rigid Russian stand.

[26]European Regulators Group for Electricity and Gas (ERGEG) *Final ERGEG study on congestion management procedures & antihording mechanisms in the European LNG terminals* (Brussels: European Regulators Group for Electricity and Gas, E10-LNG-11-03b, 2011) p. 9.

Thus, it was proved, the Russia has not enough share and power to determine natural gas prices in the German market.

Italy, the second largest importer of Russian gas, is building several new LNG terminals. Once competitively-priced LNG starts to flow it will cause a significant burden to Gazprom's Italian trading partners if the Russian company persists in maintaining its current inflexibility in pricing. While Italy, Germany, France and the UK are significant merchandise exporters to Russia, the new eastern European members do not have similar trade-flow-based bargaining leverage. They also lack adequate infrastructure access to more liquid western European natural gas markets. Brussels has recognized this problem; funds were allocated to make the existing infrastructure reverse-flow-capable and the third energy package envisaged a more rigid unbundling conditions. Futhermore member states were obliged to address free access to cross-border capacities and when necessary to build new ones. The implementation of 994/2010/EC regulation[27] basically addresses all the major security threats of pipeline supply of the eastern European internal market. The ongoing discussion on creating a common natural gas market (Gas Target Model)[28] aims to provide liquid markets across the entire territory of the EU.

The measures taken by the EU will bring results—similar to the ones that EU market liberalisation has already achieved in Western Europe, where well connected infrastructure was given—by the middle of the decade. As expected under the new regulatory and infrastructure regime, the eastern European countries will be able to address any supply interruption similar to those in 2006 and 2009, when Russia closed the taps on Ukraine. Russia learned its lesson and was particularly keen to reduce its interdependence on transit countries and avoid similar conflicts that significantly harmed its international reputation. The heavily criticized North Stream pipeline, which runs under the Baltic Sea and bypasses transit through Belarus, Ukraine or Poland, has been switched on. Poland and other CIS

[27]European Parliament and the Council: *Regulation (EU) No 994/2010 of the European Parliament and of the Council of 20 October 2010 concerning measures to safeguard security of gas supply and repealing Council Directive 2004/67/EC.*

[28]Further information about the concept at European Energy Regulators' website (www.energy-regulators.eu).

countries became less secure, as their transit bargain chip disappeared, and are no longer able to bargain on prices and delivery conditions in exchange for transit. On the other hand, although Poland is unwilling to recognize it, Warsaw gained an alternative supply route from the same source. Any new transit dispute with Ukraine or Belarus will not threaten Polish supply, providing adequate surplus capacities are available on the North Stream pipeline.

Similarly, Moscow's intention to guarantee the security and independence of its supplies from rouge transit states led it to propose the South Stream pipeline. This route runs under the Black Sea and would bypass Ukraine. South Stream is perceived by some[29] as a bluff in order to distort the development and available finances of other competitive pipeline projects that would provide alternatives to Russian supplies from the Caspian basin. The EU calls these alternative options the Southern Corridor. It consists of several different pipeline projects that partially compete for suppliers and deliver gas to different parts of the European markets. Nabucco, based on Azeri, Northern Iraqi and probably Turkmen resources, would bring natural gas to the Central European Gas Hub, in Baumgarten, Austria. This would be the first pipeline project that would aim to deliver natural gas from different fields in different countries along one major shipping route. British Petrol (BP) has recently announced an alternative pipeline along the route of Nabucco. The South-East Europe Pipeline (SEEP) advantage over Nabucco would be a significant reliance on the existing infrastructure along Nabucco's planned route, thus building only the most necessary—around 1300 km—new pipeline.[30]

The Italy-Greece-Turkey-Interconnector (ITGI) and Transadriatic-Pipeline (TAP) are competing with Nabucco (along with SEEP) for the same Azeri production at Shah Deniz II field. However, ITGI and TAP pipelines would deliver gas to Italy instead of Central Europe, with minor differences in routes.

All of the proposed Southern Corridor projects would strengthen Turkey's geopolitical position. Ankara, having gained the leverage of a

[29]Vladimir Socor, Xypaki Maria, Andrew E. Karmer.

[30]Instead of Nabucco's 3800 km, "BP plans gas pipeline to Europe from Azerbaijan," *The Financial Times* (26th September 2011).

significant transit country, will have closer relations with the EU and a stronger voice across the region. Although Iran currently struggles to meet its internal demand, in the middle term (post-2020) it can strengthen its export position and use Turkey and its new infrastructure to enter European markets. Clearly, it would need Ankara's tacit agreement, which at the moment is hard to imagine. Nevertheless, it is to be seen which direction Turkey is heading and how it is rising as a major regional and increasingly independent player in the region.

It is clear that no more than one of the above mentioned pipeline projects in the Balkans would be economically feasible, as the regional demand would not be able to absorb all those amounts that these pipelines would carry with their nominal capacity. Project development has been going on for almost a decade in the case of Nabucco, but the laying down of pipeline has still not started in any of the above mentioned projects.[31] Furthermore, Azerbaijan has not committed its Shah Deniz II production to any specific transport project. Meanwhile, the alternative Southern Corridor projects seem to be idle,[32] while Russia has intensified its efforts on the South Stream project by boosting the staffing of the project company. However, this may turn out to be insufficient for actual pipeline delivery. As time passes it is clear that none of the above-mentioned projects will come online before 2017.

By the end of the decade many new pipelines will be laid down within the European Union and some across its external borders. Challenges will no longer result from technical bottlenecks or unilateral dependence, but rather from a mix of competing producers for European and global gas markets, which could intensify regional confrontation or competition for influence. These regions can be identified as (1) Central Asia-Russia-China-(Europe), where Europe will play a small role alongside Chinese and Russian competition; (2) the

[31]The Greeks claim, that due to an earlier capacity increase at the Turkish-Greek border, they have already developed 20% of ITGI. (Costis Stambolis presentation "*A New Actor in the Caspian Energy Chain: Greece and Prospects for TAP and ITGI*" 14th July 2011, Baku, and unnamed source from Hellenic Ministry of Public Order).

[32]Lot of work have been done behind the close doors, but no visible breakthrough has been achieved in a sense that there is still lack of commitment from the producers and the construction phase has been delayed.

Caspian Basin with Azerbaijan-Iran-Turkey, where each of these countries will seek to enhance its capacity to boost its geopolitical interests by gaining strength through exports or transit.

In this pipeline competition, Central Asia can play a key role. This region has the potential to supply Europe and the Far East at the same time, a scenario that Russia clearly wants to avoid. Nevertheless, in the long run the role of the region in transatlantic supply security can increase, as it can guarantee the price link between the Atlantic and Pacific Basins. As a consequence, any turbulence in this region can lead to price shocks and global gas trade disturbances. In the future, when the number of natural gas consumers increases, avoiding global price shocks of natural gas will rise to the top of the agenda. Global market liberalization will lead to diverse and liquid markets, yet the markets are insufficient to provide security and stability in producer countries or shipping routes. The future of natural gas security will increasingly hinge on the stable and predictable pricing of natural gas.

Conclusion

The largest currently known conventional natural gas reserves reside on the axis of northwest Siberia down to Qatar (the Siberian-Persian axis consists of Russia, Iran, Qatar, and Turkmenistan[33]). While the United States is independent from the resources of this axis, the European Union views the South-Siberian-Persian axis (Azerbaijan, Turkmenistan, North Iraq) as a potential source region for its current (Norwegian, Algerian) pipeline imports, which are depleting over the midterm, and as an alternative to the currently dominant Russian supplies. As the EU taps into the region's resources, the Caspian region's geopolitical importance will increase. In global natural gas supply, Qatar's role will be balanced out by new capacities in Australia. However, for European LNG supplies Doha will remain a key supplier. Europe has not been able to extend its capabilities beyond the tools of soft power. Therefore, its increasing reliance on distant resources will require support in the framework of transatlantic cooperation.

[33]The order of these countries represents their proven reserves (BP 2011).

The unconventional natural gas production revolution not only resolved the supply security of the United States, it significantly eased pressure on European supply security challenges. The amounts diverted from the U.S. to EU markets significantly diminished earlier European supply concerns, and led to diversified supplies. Currently the EU's is addressing internal regulatory and infrastructure bottlenecks that will enable Brussels to deal more effectively with threats that may be related to Russian-origin natural gas supplies.

Chapter Twelve

Towards a New Balance with Russia? Russian Energy Challenges and the West

András Deák

Russia is the biggest hydrocarbon producer and exporter in the world. In 2010 combined Russian oil and gas production was above the U.S. level by 15%, Russian exports exceeded those of Saudi Arabia by more than 50%. At the same time Russian fixed capital costs are high, and the demand situation is relatively bad in global comparison. The western Siberian and Arctic regions belong to the worst production sites of the planet from a climatic, geographic and infrastructural point of view. In the 1980s construction costs were four to eight times higher in western Siberia than in temperate Russian locations; labor costs were two to seven times higher, and infrastructure development costs were sometimes ten times higher.[1] Moreover, western Siberia is halfway between Berlin and Tokyo, and all across this distance both oil and gas have to be transported by pipelines. According to some estimates 70% of all investment costs in the gas industry were spent on transportation during the 1980s, and these numbers seem to be close to current proportions.[2] During the intensive phase of development of these regions, every four to five weeks a pipeline network equal in length to the Alaskan pipeline was laid down.[3] Due to pipeline con-

[1] For a good overview of the climatic determinants: Victor L. Mote, "Environmental Constraints to the Economic Development of Siberia," in: R. G. Jensen; T. Shabad; A. W. Wright eds., *Soviet Natural Resources in the World Economy* (Chicago, London: University of Chicago Press, 1983), pp. 15-72.

[2] Thane Gustafson, *Crisis amid Plenty—The Politics of Soviet Energy under Brezhnev and Gorbachev* (New Jersey: Princeton University Press 1989), p. 149. Gazprom puts the high-pressure transportation network at 50% of its basic gas branch value. Gazprom Annual Report—2010, Available at http://gazprom.ru/f/posts/42/228071/gazprom-annual-report-2010-rus.pdf p. 57.

[3] Mote, op cit., p. 38.

nections, export options are locked in at the initial stage of development. Russia shares not only the global producer risks, but its exports exclusively depend on European, and in some cases on some national, market developments.

The factors determining Russian energy policy to a large extent, especially in its external aspects, are:

(1) Russia is a high fixed capital cost conventional oil and gas producer.

(2) Demand security considerations prevail in Russian energy thinking.

These two features make Russian energy production vulnerable. The paradox is that despite all these handicaps, Russia could still become by far the largest global energy exporter. According to classical economic theory, low-cost areas with better marketing opportunities and higher profits should take a bigger share of the market and increase their drilling faster, than high-cost producers. Russian exports have been showing the opposite in the past two decades. This also means that Russian production prospects are becoming dependent not only on domestic policies and the industrial technological progress, but also on existing patterns of the global marketplace. This study tries to explore Russian energy policy efforts to tackle these risks and decrease their potential impacts.

A Dual Export Strategy—Gas and Oil

Russia has a peculiar external energy strategy, as it exports both gas and oil to foreign markets on a large scale. Most producers in a similar situation focus only on oil, since it is more transportable and has an energy density four times bigger than gas. Russia is not fully an exemption from this general rule: crude oil and oil product export revenues reached $206.2 billion, while Russia earned only $47.7 billion from gas exports in 2010.[4] However, natural gas still has a remarkable share, and according to the Russian Energy Strategy, it is definitely

[4] Source: Russian Central Bank, Available at http://www.cbr.ru/eng/statistics/?Prtid=svs. Traditionally these data slightly differs from those of the Custom Service.

gas exports that should grow by roughly 40% in the next 20 years.[5] Russia has a dual export strategy.

Accordingly, gas and oil markets will be separated in the following sections. They differ both technologically, institutionally, with respect to global market patterns and Russian regulation and internal situation. The main differences can be summarized as follows:

(1) Oil demand is more inelastic mainly due to the lack of substitute fuels in transportation. More than 60% of oil consumption comes from the transportation sector. Further efficiency gains and potential substitutions in chemical and electricity sectors are scarce. This results in higher price volatility, consequently a bigger potential for cartelization than in gas sector. Developing export strategies for oil is an easier task than for gas. The situation may change only by breaking oil's monopoly in the transportation sector. Understandably Gazprom officials are eagerly looking for such a technological switch.[6]

(2) The oil market has a global price setting mechanism in the form of OPEC, whereas the gas market has not. Even though the oil market is a more convenient place for price setting, gas markets have been globalizing rapidly mainly due to the rise of LNG trade. This also opens up a certain potential for institutionalization of global producer cooperation, and some early attempts in the form of GECF (Gas Exporting Countries Forum) can be observed. Russia traditionally pursues a free-rider policy towards OPEC: contacts are rather cautious and have not reached a strategic level. Production cuts were supported by Russia only sporadically and rather rhetorically, if at all. However, since Russia has become a major exporter during the last couple of years, its dependence and interest in global oil price levels has increased significantly. It can be stated, that the immanent Russian

---

[5] The Strategy envisages stagnating crude oil exports (not providing a numerical estimate for oil product exports), while gas exports prognosed to grow from 243 bcm in 2008 to the range of 349-368 bcm in 2030. Source: Energy Strategy of Russia for the period up to 2030. Available at http://www.energystrategy.ru/projects/docs/ES-2030_%28Eng%29.pdf p. 136.

[6] Apart from smaller vehicles, like automobiles that can be fuelled both by electricity or CNG, there is a German-Russian pioneer project to put heavy truck traffic on LNG in Europe. Interview with Sergey Komlev, Gazprom, Head of Contract Structuring and Price Formation Department, Moscow, November 24, 2010.

reliance on OPEC has grown and the cartel's policy is not a neutral issue in Moscow anymore. But free-riding is not an option in the gas markets yet. There is no cushion against recession and Russia has to adjust both its production and pricing to market dynamics.

The 2008-2009 economic crisis clearly demonstrated different impacts on Russian oil and gas exports. In the oil market Russia did not have to join the OPEC production cut and could increase both its crude oil and product exports. Accordingly exporters suffered only the consequences of the price collapse in 2008-2009, cushioned by the OPEC and even could slightly offset its impacts by increasing sales. In the gas sector Gazprom faced both the reduction of its exports prices, and also a 15% decrease in its export volumes in 2009. While Russian oil export revenues dropped by 14% between 2008 and 2010, Gazprom suffered a 31% drop in the same period and recovery seems to be more sluggish.[7] The reason for the difference between the two outcomes is that the oil market has a price-setting mechanism and Russia enjoys its benefits without the commitments. On the gas market there is no global price-setting mechanism and Russia has to suffer all the consequences of a recession. Supporting cartelization and being a member of it is somewhere between the two strategies.

(3) Russia has a resource bottleneck in the oil production, while gas reserves are abundant. Even though Russia has the biggest proved conventional oil reserves outside OPEC, and according to a 1998 U.S. Geological Survey the biggest undiscovered reservoir,[8] due to its high production level and vast area where these fields are located, reserve replacement remains a problem. Reserves in particular regions and their quality put a constraint on future developments, questioning the mid-term sustainability of current levels and future increases. Unlike oil, the abundance of natural gas deposits offer a high number of different development paths for producers. Gazprom alone has half a dozen gas fields with the size of each equal to Europe's biggest forma-

[7] Even if these numbers are not fully comparable because of the different pricing in the two sectors. Source: Gazprom in numbers 2006-2010, Available at http://gazprom.ru/f/posts/42/228071/gazprom-reference-figures-2006-2010-rus.pdf.

[8] Referenced by John D. Grace, *Russian Oil Supply—Performance and Prospects* (Oxford: Oxford Institute for Energy Studies, 2005), p. 182.

tion, the Groningen field.[9] Having one quarter of global reserves, gas output can be increased if corporate ambitions are present and market conditions allow.

(4) The two industries have different market and policy patterns in Russia itself. Oil industry has an oligopolistic structure and a solid export orientation. There is a consensus at the corporate and governmental level that oil exports are crucial for Russian welfare and exports should have the priority. Unlike oil, 70% of gas production goes for domestic consumption, covering more than half of Russia's primary energy demand. Cheap gas is a major social benefit for the population and an important subvention for a high number of influential lobbies in the processing sector. The relationship is further complicated, because prices and market mechanisms do not play a decisive role in balancing between domestic consumption and exports. It is the interplay of many corporate and governmental actors that shape the regulation and determine the form and scale of cross-subsidies between the two. Theoretically, given the large reserves, Gazprom could separate the two segments and pursue an export policy independent from domestic considerations. But Gazprom's export monopoly has been questioned by many other lobbies, and the company cannot be sure about the future distribution of its export revenues. Thus, prospects for gas exports are surrounded not only by domestic supply tasks, but also by internal policy and political uncertainties.

Despite these differences, there is at least one major factor identical in the two industries: exports are much more profitable than domestic sales. Only export rents are divided differently. The government takes much of the export revenues in the oil industry, while gas exports have a lower export custom and royalty level combined with much lower domestic prices. However, on the macroeconomic level the rationale of directing limited financial and political resources to exports is indisputable. All the rest is an internal Russian divide. The question is how to increase exports further, without increasing external vulnerability stemming from high fixed costs and high demand dependency. There are three possible ways to do so:

[9] Johnathan P. Stern, *The Future of Russian Gas and Gazprom* (Oxford: Oxford Institute for Energy Studies, 2005), p. 9.

1. Diversification of risks at the level of transit, export destinations and, in the case of oil, at the level of products (rebalancing crude and oil product exports).
2. Increasing energy efficiency domestically, thus decreasing development needs and average capital costs.
3. Establishing corporate and/or industrial synergies in foreign markets in order to decrease demand risks. This may mean acquiring downstream assets in export markets, letting foreign companies into the Russian industry or strengthening price setting mechanisms and cooperation with other producers.

The current Russian external energy policy combines all these potential solutions. It would be difficult to pick one and qualify it as a priority for Moscow. There is no final engagement and it is not sure that there will be any in the upcoming years. The following sections offer an overview and evaluate these policies, respectively, on the oil and the gas markets.

Has Russia Becoming a "Normal" Oil Producer?

Most of the challenges of Russian external oil policy are to be found in Russia itself. Due to the substantial call from global, primarily Far Eastern markets and the existence of OPEC, effectively capable to cushion oil demand and price collapses, the mid-term external environment seems to be favoring even high-price exporters like Russia. Moscow and Russian companies can set export targets relatively freely, both for crude oil and oil products. Russia efficiently used the emerging sellers' market of the 2000s: its incremental production covered half of global market growth between 2000 and 2010.[10] The basis of this export offensive was primarily the expansion of the export infrastructure and the refurbishment of already existing fields and, to a lesser extent, some selected new field developments in western Siberia, such as Priobskoye or Vankor (the latter formally located in Eastern Siberia, but connected to the western Siberian infrastructure).

[10]Increasing its share in global production from 8.7% to 12.5%.

Maintaining a production level beyond 10 million bpd is extremely difficult, especially with Russian capital costs and a reserves-to-production ratio of 20.6. Much of the production increase has happened between 1999 and 2006, since then output has been growing at a much lower rate. The only published production target is in the 2009 Energy Strategy, which does not foresee further export increases; only domestic consumption is expected to grow in the range from 0.8 to 1.5 million bpd until 2030.[11] Accordingly, Russia appears to be unwilling to further increase its exports. On the basis of past underestimations, this target has to be taken with certain caution. At the same time Moscow will need increased efforts to sustain or even increase its production. It is telling that Moscow has launched a series of tax reforms in the last couple of years in order to overcome potential constraints at investments into production capacities. Since 2008 there has been a shift towards a more pin-point taxation policy with differentiated levels, benefiting new, mainly eastern Siberian, off-shore deposits and smaller fields.[12]

All these shifts in taxation could provide huge additions. James Henderson has examined company data and has projected 1.5 million bpd of incremental supply from eastern Siberia by 2020.[13] More focus on smaller and less extractable deposits may contribute to production levels even more in the future. Having a mature, even declining reserve base with infrastructure on the spot, turning to formerly ignored smaller deposits and strengthening post-mature production strategies is almost a must if Russia wants to level off its oil revenues. However, corporate structure, philosophy and governmental taxation have not been fully adjusted to these tasks. It is very likely that new additions will be sufficient to flatten the production until 2015, but it is unclear where the incremental supplies will come from in the second half of the decade.[14] The current measures will have to provide their benefits until then.

[11]Energy Strategy of Russia for the period up to 2030, p. 135.

[12]S. Sinelnikov-Murilyev; A. Radigin; N. Glavackaya eds., *Rossiyskaya ekonomika v 2010 godu. Tendencii i perspektivi* (Moscow: Institut Gaidara, 2011), pp. 267-285.

[13]James Henderson, *The Strategic Implications of Russia's Eastern Oil Resources* (Oxford: Oxford Institute for Energy Studies, 2010), p. 66.

[14]Russia may have huge reserves in the Arctic off-shore regions, but since these fields are unexplored and are at the edge of technical capabilities, they cannot contribute until 2020-2025 to the overall production level.

At the same time, Transneft's (the Russian oil pipeline monopoly's) export pipeline capacity expansion shows a slightly different picture. According to its current plans, total Russian export pipeline capacity is going to be raised by 3.3 million bpd until 2015.[15] This capacity is unlikely to be fully utilized, even if Caspian Pipeline Consortium expansion is meant for increasing primarily Central Asian crude oil transit to Russian Black Sea ports, and not for Russian exports itself. Thus, Transneft will be able to allocate idle transport capacity for the first time since the Soviet split-up. It is reasonable to think that shipments will be diverted from inconvenient and expensive trade routes to new pipelines. This may negatively affect the Druzhba-pipeline and the Black Sea terminals.[16] The high-level of idle capacity also suggests that construction of new crude oil export pipelines is extremely unlikely even after 2015, but some small-scale additions towards the Pacific or China are possible.

Another factor that can influence crude oil exports is oil product consumption and exports. The Russian government consciously supports increases in higher value added oil product exports, while companies do so with less enthusiasm. The volume of total oil product exports more than doubled from 1.26 million bpd in 2000 to 2.67 million bpd in 2010.[17] The basic problem is that much of this increase came from low-standard oil products (primarily fuel oil).[18] Apart from taxation, which was adjusted recently in order to further promote exports, three factors may limit further increases: (1) the booming domestic demand of lighter products due to motorization; (2) structural problems in refining; (3) the low capacity of the export infrastructure, primarily that of product pipelines.

---

[15]Capacities under construction or investment decision taken. Source: Transneft, available at http://transneft.ru/projects/119/.

[16]Even if decrease in Druzhba exports may cause some problems in Central Europe, regional refineries have alternative supply options. *Study on the Technical Aspects of Variable Use of Oil Pipelines—Coming into the EU from Third Countries*, Available at http://ec.europa.eu/energy/oil/studies/doc/2010_reporting_technical_aspects.pdf.

[17]Source: Russian Central Bank, Available at http://www.cbr.ru/eng/statistics/print.aspx?file=credit_statistics/oil_products_e.htm.

[18]In 2008 52 percent of total oil product exports, almost 1 million bpd, was fuel oil. V.V Busuev; A. M. Mastepanov; A.I. Gromov eds., *Toplivno-energeticheskiy kompleks Rossii: 2000-2008 gg.* (Moscow: Institut Energeticheskoy Strategii, Energiya, 2009) p. 168.

Domestic demand due to fast motorization is the major driver in the modernization of refining. Between 1995 and 2009 the number of cars grew from 14.2 million to 33.1 million,[19] the majority of which were Western-produced. This trend may continue in coming years: the current car per capita ratio is less than half of that in Hungary and less than one fifth of that in the U.S. At the same time fuel oil consumption has fallen by more than half in a decade, due to changing electricity generation and heating patterns. These latter volumes have been redirected to exports, because domestic refining was inadequate to process it. Accordingly, the low depth of refining (around 70% on average) is a major bottleneck.

The modernization of the refinery sector is the basic efficiency reserve that the Russian oil industry currently has, and probably the most reasonable growth path in coming years. Better utilization of fuel oil may help to meet growing domestic gasoline demand and provide incremental diesel yields for exports. This may ease the need for further investments on the supply side and help to maintain current export levels. However, the question remains open whether oil companies will have the financial resources to maintain production levels and increase refinery complexity simultaneously. Despite government efforts and tax preferences, companies must allocate significantly more capital than before to invest into both upstream and mid-stream. Due to low refinery margins worldwide, companies will need further stimulus than just abundance of fuel oil and cheap domestic gas prices to opt for comprehensive refinery modernization.

All in all, the Russian oil industry has been completing two decades of post-Soviet transformation. During these 20 years oil exports from Russia have increased in volume, became more diversified in terms of destination and more direct in terms of transit. In the next few years they may become even more diverse in terms of export products. Russia has managed to redirect its production capacities to foreign markets, has switched domestic electricity and heating systems from oil to other fuels, and has adjusted its export infrastructure to the new realities. Through these fulfilled tasks, much of the transformative and efficiency agenda seems to be exhausted. Russia has become a "normal

[19]Source: Russian Statistical Office, available at http://www.gks.ru/wps/wcm/connect/rosstat/rosstatsite.eng/figures/transport/.

producer" whose exports performance depends overwhelmingly on investments in new production. Since these gains are more expensive and challenging, Russia has turned greater attention to another, more complex export market, namely that of natural gas.

Who Needs Russian Gas?

Unlike the oil industry, the Russian gas industry has not transformed itself into a full-fledged export industry. The bulk of production is for domestic use; Gazprom exports almost exclusively to European countries (including Turkey and western CIS states) and still tries to manage its dependence on some transit countries. But the basic difference from oil exports is that prospects on external gas markets are much less clear and promising. Until 2008 Gazprom has had a relatively favorable situation, based on the combination of depleting reserves in Europe, global climate policies usually favoring gas consumption, and solid economic growth in its main markets. Since then, the U.S. gas shale revolution and its potential spillover to other regions overshadow global prospects. Relative stagnation and the bleak economic outlook in western Europe limit potential electricity consumption growth. In such an environment, LNG suppliers may pose a longer-term competitive constraint on the market than expected. Far Eastern, primarily Chinese, exports depend on some policy decisions in specific capitals. While some years ago Gazprom had a solid strategy with an exclusive focus on future export growth, these plans have to be at least reconsidered in the light of the new realities.

Who needs more Russian gas and under what conditions? Before the crisis Gazprom had a highly ambitious production target for 2020. Instead of 580-590 bcm as stated in the Energy Strategy, the government instructed Gazprom to raise output to 670 bcm by 2020.[20] This target was calculated on the basis of a growing domestic and European demand, with the launch of Far Eastern exports and Gazprom entering the LNG market. It is almost certain that most of these assumptions have been washed away by the economic crisis and the

[20]Alexander Ananenkov, the senior manager in charge of production at Gazprom set Gazprom's maximal potential output capacity at 900 bcm/year in 2007 (in 2007 it was 548.6 bcm). Available at http://gazprom.ru/press/conference/2007/1406/.

new realities. Gazprom's sales did not reach 508.6 bcm in 2010, more than a 9% drop in comparison with the peak output in 2006. The monopoly has significantly changed its 2009 investment plans and there is no sign of returning to its propositions. Gazprom has drastically decreased its imports from Turkmenistan since 2009, without any agreement on a potential renewal of increased supplies.[21] The economic crisis appears to be a major milestone in Gazprom's strategy. Thus, it is very important to look at Gazprom plans prior to 2008 with some caution.

At the same time Gazprom has eased its production and transportation bottleneck by entering the Yamal-Peninsula and constructing the Nord Stream pipeline. According to Russian reserve classification the region has 16 trillion cubic meters of proved and probable gas deposits. By 2015 Gazprom would like to produce 115 bcm in the first phase of the Bovanenko-field. This was a necessity if Gazprom wanted to balance its production. However, once these reserves have been tapped, they enhance a relatively flexible production policy for the monopoly. In the coming years it will have an easy access to other Yamal fields if needed, or can boost production by developing smaller deposits in the old Western Siberian regions. Gazprom will have both the infrastructure and the production base to increase exports easily and swiftly. Gazprom has a number of options to form its "future supply roadmap,"[22] and one of the main variables remains its exports to external markets. Given the large investments into infrastructure and the possibility of increasing export volumes, there is a permanent temptation to increase sales at the export markets, even if by concessions on contractual terms.

---

[21]The sharp decrease in Central Asian imports (around 40 bcm in 2010) very much challenges the credibility of South Stream with its previously announced capacity of 63 bcm. Even with Azeri and Southern Russia supplies it would be difficult to free enough gas in the region for South Stream and Blue Stream. Gazprom should bring more gas from Western Siberia to this roundabout way, an unusual option even for the gas monopoly. Maybe it is no accident that statements on future South Stream capacity have been rather scarce since 2008.

[22]Jonathan Stern, *Future Gas Production in Russia: Is the Concern about Lack of Investment Justified?* (Oxford: Oxford Institute for Energy Studies, 2009) p. 3. Available at http://www.oxfordenergy.org/wpcms/wp-content/uploads/2010/11/NG35-FutureGasProductioninRussiaIsTheConcernAboutLackofInvestmentJustified-JonathanStern-2009.pdf.

Gazprom senior officials traditionally refer to the Russian domestic market as the most promising one. Their optimism is based on the government policy that aims for equalizing profitability of domestic industrial and export market prices (net netback prices) by 2014.[23] This would mean higher prices, competitive with those of the exports for domestic industries and likely a more modest increase for the population. Despite the revolutionary potential of such a development, it has to be said that initially the deadline was set earlier and tariff increases still lag much behind the schedule.[24] Consequently, domestic prices seem to be out of Gazprom's control and therefore references to the emerging Russian market as a major source of income assume a good deal of optimism.[25] This also assumes that demand efficiency gains coming from more cost-reflected pricing may remain moderate and the internal market will remain huge, but relatively underpriced.

Gazprom has also been losing its share in the Russian domestic market, showing a good deal of negligence or inability to preserve its positions. The share of oil companies and the so-called "independent" gas producers has been growing steadily since 1998 and already reached 23-24% of total consumption.[26] Even if Gazprom, the owner of the high-pressure pipeline system, can get some of the benefits of others' production, it would be difficult to qualify it anymore as a monopoly player on the internal market. There is no single reason behind these trends. On the one hand, oil companies and the few

---

[23] Alexey Miller usually refers to this policy when he argues that soon Gazprom will have "two European markets" in Russia.

[24] Average domestic wholesaler industrial prices nominated in USD have grown from 19.64 percent to 27.28 percent of the gross average non-CIS exports price level between 2007 and 2010. Between 2003 and 2010 average domestic (industrial and residential) gas prices grew from 689 RUB/mcm to 2392 RUB/mcm, only a 1.71 times increase in real terms. During the same period average gross export prices to non-CIS countries have risen from 131.6 USD/mcm to 301.8 USD/mcm, a 1.93 times increase in real terms. It can be stated that the "equalization" policy has closed the gap only very modestly until now. Author's own calculations, data from Gazprom and Russian Central Bank.

[25] Despite the 54.5 percent share of the internal sales, Gazprom earned (after taxation) only 28% of its revenues from domestic markets in 2010, slightly more than in 2006 (25%).

[26] For a good overview: James Henderson, *Non-Gazprom Gas Produces in Russia* (Oxford: Oxford Institute for Energy Studies, 2010).

independent producers are relatively well entrenched both in the market and behind the Kremlin's doors. It would be difficult to qualify Gazprom's current standing as guaranteed, in light of the political infighting of the Russian leadership. On the other hand, Gazprom prefers exports to the domestic market. Having an export monopoly in its hands, it has the opportunity to swap from internal supplies to external, giving up some of its positions to other producers. "Independent" production gives Gazprom a comfortable supply buffer in the domestic market.

Given these inflexibilities, despite its dominant share, domestic market conditions largely remain externalities for Gazprom. It cannot formulate an independent price policy because of the administrative pricing and the obvious counter-interests of price hikes, and can only increase its sales volume with certain caution. The only sure option for increasing its profits on the domestic market is expansion along the value chain, vertical integration and utilizing industrial synergies. During the last couple of years Gazprom heavily invested into related Russian energy assets like oil production, petrochemical industry or electricity generation. Today the gas industry only has a 59% share in Gazprom's total net revenues.[27]

Paradoxically, Gazprom can set its policy more freely on export markets than at home. It can autonomously decide about its pricing and export volumes. While Gazprom is constrained by administrative pricing and supply obligations on the Russian internal market, only competition—and to a lesser extent the EU and national regulators—may limit its leverage in Europe. After the crisis Gazprom pursued a "wait and see" policy, and strategic decisions about adjustments were postponed until the post-recovery period. This was a reasonable strategy and Gazprom's export revenue did not decrease as much as those of other, traditional suppliers which gave price discounts (like Statoil). However, maintaining this policy poses a threat to Gazprom's market share in the longer term. Three years after the crisis the mid-term prospects are still bleak, while Gazprom has lost approximately 20 billion bcm of European exports. For European importers Gazprom's contractual inflexibilities, the relatively high level of minimal takeover

[27]Gazprom in figures 2006-2010. Table "Gas Sales," available at http://gazprom.com/about/management/shareholders/2011/.

amounts and full-fledged oil-link pricing are inconvenient and uncompetitive. Companies like Edison or E.ON started or threatened to launch legal procedures in order to quit their existing contracts. The long-term traditional contractual system turned from Gazprom's demand security guarantee into a threat to its export market share. Without changing the conditions importers will probably not prolong their expired contracts and will turn to other sources.

This is a tough choice for Moscow. If they start modifying their export contracts, there will be hardly any way back. Gazprom will have to loosen up its "take-or-pay" system and include spot-pricing into its pricing formulas in almost all contracts. This may be a losing strategy if European economies in the end cope with the crisis or gas demand continues growing in the region. But if it insists on the existing patterns it may lose its credibility and European companies would minimize their sales from Russian sources. There are more arguments for the former strategy. Gazprom invested heavily both into production and transportation before the crisis, so it is in its best interest to utilize the new capacities. Gazprom also tried to enter other, North-West-European markets before the crisis and launched an aggressive price campaign to get a reasonable share. This very well demonstrates its flexibility in contractual issues—if the conditions force it to do so. In the end, it has no other export markets than Europe. If it had a considerable share in the Far Eastern markets, it could change its investment or export policy. But this depends very much on Chinese import policy, and Beijing seems to be unwilling to consent to European prices.

For Gazprom a slow and cautious process of contractual concessions could guarantee its existing market share. Thus, it is no surprise that both non-long-term (it is difficult to define what "long-term" means for Gazprom) and spot-priced sales increased to 7% within total Gazprom exports, and that the company raised its direct sales to 5 bcm in the EU in the midst of the economic crisis. There is an extensive discussion on lower take-or-pay obligations in the contracts and about incorporation of spot-prices into the price formulas. Gazprom has to adjust its policies to realities in Europe and this very much questions the prospects of a Russian gas export offensive until 2015. Export markets seem to be constrained and Gazprom has not found the market to which to sell its gas in the near future.

Shaping Relations—Russian Energy Exports and the West

Traditionally, Russian companies provide access to their reserves in two cases: for technology and assets in consumer countries. Having a complex production background and high level of demand security requirements, an existing but outdated machine-building and metalurgic sector, the Russian energy industry shows great affinity in these questions. This makes the Russian energy sector more accessible to Western companies than many of its counterparts in the Middle East. Despite past limitations to foreign ownership, most Western companies have significant production assets in Russia, even in the gas industry. BP has a 50% share in the third biggest Russian oil major, despite rude takeovers in their Sakhalin-projects Shell and Exxon preserved their assets there, E.ON and Wintershall have gas production assets at the Yuzhno-Russkoye field in the framework of the Nord Stream deal.

However, the underlying considerations are different in oil and gas industries. The Russian oil industry needs primarily technology, while the gas sector is short on markets. Having a resource bottleneck, it will be difficult for the Russian oil industry to sustain its production levels without making the first steps in the Arctic off-shore region or tapping the hard-to-extract on-shore deposits. Any development path without a massive involvement of foreign technology in these two fields seems to be improbable. While Gazprom could postpone the exploration of the Arctic Stokman-field at the first signs of an economic crisis and rely on its huge, traditional reserve base, this is hardly an option for many oil companies. The Arctic off-shore frontier is an opportunity for Gazprom, but a must for Rosneft.

The Russian continental shelf is still underexplored, but even at the current stage it gives more than one-fifth of total reserves.[28] Almost 70% of these fields are located in three Arctic seas (the Kara-, the Okhotsk- and the Barents-sea). The Russian Energy Strategy qualifies this region as the major source of compensation for Western Siberian depletion.[29] It is difficult to quantify this "production call" as there are different projections for Western Siberian decline. According to

[28]According to Russian reserve classification.

[29]Energy Strategy of Russia, p. 60.

announced but not approved plans, regional oil transport capacity can be increased to 2.2 million bpd by 2015 from 0.67 million bpd in 2008—a strong overestimate for potential needs (understandably these terminals would transport at the beginning primarily on-shore oil from the region).[30] Russia has an extraordinary expertise in Arctic production, gained during the previous 40 years. At the same time Russian know-how and technology with regard to off-shore is very limited. Gazprom reportedly faced serious difficulties at its Prirazlomnoye oil field, a pioneer Arctic off-shore project, which is just 60 km from the shore in shallow waters. Establishing an efficient infrastructure and extracting other fields further in the high seas will be a challenge even for the most trained personnel and superb technology. It is highly unlikely that the Russian oil industry will be able to cope with this challenge alone. Moscow will have to cooperate with those few Western majors which gained critical expertise in Arctic off-shore drilling. Thus, the recent Exxon-Rosneft deal on Arctic and Black Sea cooperation may be a first step in entering the Russian Arctic continental shelf.

This will be a difficult choice for Kremlin politicians. Apart from the huge tax-exemption, provided for investors on the continental shelf, they will have to provide concessions for Western companies. Not surprisingly, the discussion on easing the current legislation on strategic industries and reserves has a low-profile and senior Russian leaders barely make conciliatory comments on these issues. Industry is still preoccupied with eastern Siberia and new additions from there provide few years flexibility in choosing the next locations for development. Nonetheless, time is not on the side of Russian companies and new fields in other Arctic regions; other countries may create a serious technology capacity bottleneck for the years to come.

At the same time the issue is not neutral for Western countries either. A decline in Russian exports after 2015 could significantly contribute to their dependency on OPEC. More than one-third of non-

---

[30]The prognosis was provided by the Russian Ministry of Economic Development. Alexei Bambulyak; Bjørn Frantzen, *Oil transport from the Russian part of the Barents Region. Status per January 2009* (The Norwegian Barents Secreteriat, Akvaplan-Niva, 2009) p. 38. Indra Overland estimates 2.2 million bpd oil and 160 billion bcm gas production from Arctic off-shore in 2030. Indra Overland, "Kooperation statt Konfrontation," *Osteuropa* (2011/2-3 "Logbuch Arktis"), pp. 129-143, p 131.

OECD, non-OPEC production comes from Russia; these amounts cannot be easily substituted from other locations. It was only 20 years ago when the world witnessed—despite the Soviet split-up and the collapse of the planned economy—a 5 million bpd drop in Russian supplies.[31] Even a significantly smaller drop would put tremendous pressure on oil markets now. This provides some leverage for Moscow. Extensive resource nationalism or the failure of renewal of its production assets would influence global price levels. During the past decade Russian oil exports have reached a level which makes them impossible to ignore from a global oil governance point of view anymore. This is an important point of consideration, even if Russian decision-makers often overestimate their own significance.

The gas industry is a more complicated issue. Technology is not such an urgent necessity in Russian gas production, the sector has more security considerations and governments have a bigger say through regulation. The Russia-West dialogue is based on Gazprom's ambitions to increase its exports on the one hand, and the EU's policies to increase transparency and efficiency of the Single Market and central European national efforts to decrease their dependence on Russian gas imports on the other. The United States is less involved in these issues, due to its self-sufficiency and oil-focused energy policy. Lengthy and complex coordination among European nation states causes understandable frustration in Moscow, while Russian energy philosophy appears to be an unchangeable and imperialistic dictate for many European decision makers.

Gazprom's desire to get closer to European consumers is understandable. Having a long and expensive transport infrastructure to the European markets, it would like to lock in its markets and be sure of efficient returns of its past investments. While this was more or less guaranteed in the mid-2000s during the time of steady growth, the environment became more risky due to increased competition on these markets and the EU's third energy package after 2008. What is more, Gazprom gets financial benefits from entering European markets. Unlike in the oil industry, where Russian taxation is based on globally benchmarked prices, gas export customs are calculated on the

[31]OPEC cut production in 2008-2009 by a comparable volume to balance oil market prices in the midst of economic crisis.

basis of contractual sales at border prices. But Gazprom can use "transfer-pricing," giving price concessions from border prices if in a joint venture or alone can sell its gas to consumers directly, benefiting from the high margins on European internal markets. Swift cooperation with Germany and Finland resulted in significantly lower border prices than in Poland or Italy where Gazprom's entry was minimized or excluded.[32] Vertical integration and buying assets in petrochemical and electricity generation was a successful policy in the Russian domestic markets to offset negative consequences of low administrative pricing on gas. There is no reason not to attempt a similar policy on the export markets, even if political resistance and coordination with local majors make this strategy a more complex task.

All these factors make a more accommodating Russian gas policy in Europe probable. Gazprom will likely shift its policy from the current price-based strategy towards a more volume-optimization behavior, defending its existing market shares. Due to segmented European markets, these benefits may less affect central European markets than the competitive western European ones. Conversely, in the post-Soviet space Moscow has been collecting the fruits of its past conflicts. Due to relatively high price levels and high consumption, these countries will have to give further concessions, like joining the Customs Union or letting the Russian companies to further buy their gas and oil assets. Ukraine could withstand these efforts alone, if it continues to cut back its gas imports further and could boost domestic gas production by letting foreign companies to its reserves.

Both of these trends may cause political tensions in the years to come. Changing contractual patterns and accommodating to the EU's third energy package is a painful process for Gazprom. Building up verticality in the post-Soviet space has led in the past to cutoffs and transit wars, which cannot be excluded even in the future. However, since Gazprom will need additional markets, this makes the EU-Russia gas dialogue more market-based than before. Politics cannot help too much in a situation with increased competition. Thus, the current buyers' market is a window of opportunity for EU governments and

---

[32] This income usually lands in Swiss or other offshore companies. This also complicates comparing Gazprom prices or controlling its income for Brussels and the Russian government.

the Commission to pursue a more independent policy towards the single market, help central Europe to diversify its supplies and even for Ukraine to cope with its energy problems. Given the current supply situation, all these efforts can be done within a more balanced relationship with Gazprom.

Summary

The Russian energy industry has a strong autarkic perspective. Apart from its early 19th century beginnings, it has been developing on a solid national basis; cooperation with foreign majors has been much more limited than has been the case with other producers. Soviet energy has been produced exclusively by national champions since 1917, limiting cooperation with regard to the import of technology and exporting oil and gas. Moreover, Russia became a great power long before it started its energy exports. Unlike Saudi Arabia or Iran, for which oil was almost the single reason to become a foreign policy actor, Russia has to integrate its global energy significance into a ready set of external relations and perceptions. This makes Russia a peculiar actor. It does not have the sometimes negative experience of having foreign concessions on its soil, but it has a high level of expectations regarding its status and leverage on world general and energy matters.

Russia is now approaching the limits of its autarkic industry patterns. It will have to establish a new, more cooperative policy in oil production if it wants to secure its production level in the long term. It has to show a more interactive behavior with gas exports if it wants to keep its markets in the midterm. Thirty years ago similar constraints and the failure of the Soviet leadership to manage them played a considerable part in the collapse of oil production and in the fall of the Soviet Union. The situation has changed since then. The Cold War is over, but Russian leaders remember and fear a similar outcome, while energy exports play a crucial role in current Russian stability. All this provides a chance to set the fundaments for a moderate, if often uneasy, partnership with Russia in the field of energy.

Chapter Thirteen

Transatlantic Energy Security and Ukraine: Politics, Corruption and National Interests

Taras Kuzio

Ukrainian energy policies during the last two decades of Ukrainian independence are closely tied to the country's evolving domestic politics and foreign policy. Four Ukrainian presidents[1] and twelve (of fifteen) governments have not dismantled the system in place, which draws high rents from the energy sector. As Chatham House's James Sherr states, no Ukrainian government has sought to break the pattern of dependency, opacity, rent seeking and preferential pricing, since any such effort would have broken the close ties between big business and power.[2]

The three governments that were the exception to this rule were Viktor Yushchenko in 2000-2001, where Yulia Tymoshenko was Deputy Prime Minister with responsibility for energy; and two governments headed by Prime Minister Tymoshenko in 2005 and 2007-2010. The Yushchenko government cancelled 250 government resolutions that provided subsidies to companies on energy, many of whom operated in Ukraine's large shadow economy.[3] The anti-corruption crusader Prime Minister Yushchenko proved to be very different to

---

[1] Leonid Kravchuk (1991-1994), Leonid Kuchma (1994-2004), Viktor Yushchenko (2005-2010) and Viktor Yanukovych (2010-).

[2] James Sherr, *The Mortgaging of Ukraine's Independence* (London: Royal Institute International Affairs, August 2010).

[3] Oleh Rybachuk, head of the presidential secretariat in 2005-2006, is cited in "17 myttyevostey vesniy 2001-ho," by *Istorychna Pravda*, 29 April 2011. Ukraine's shadow economy is estimated by the government and World Bank to account for 40-50 percent of GDP. See Taras Kuzio, "Political Culture and Democracy: Ukraine as an Immobile State," *East European Politics and Society*, vol.25, no.1 (February 2011), pp. 88-113.

President Yushchenko, who was a defender of opaque gas intermediaries. President Yushchenko blocked Tymoshenko's attempts to remove the gas intermediary RosUkrEnergo.[4]

Ukraine's energy sector has been the source of the greatest degree of corruption in the country and the extraction of rents from this sector have taken priority over the country's energy independence, national security and efficiency. The declared national interest of energy independence from Russia has often been defined as a key strategic goal in Ukrainian legislation and by experts, but has never become a priority for Ukrainian leaders.

With short-term frames of reference, Ukraine's elites have preferred to cooperate with Russia in opaque gas schemes rather than investing in domestic oil and gas production or reducing gas consumption through greater energy efficiency. Ukraine, which was an exporter of gas until the 1970s, has sufficient reserves to produce up to 80% of its gas needs, up from 20% currently. But investment in greater domestic production would require Ukrainian leaders with vision who are ready to implement market economic reforms and prioritize the national interest. Such leaders have yet to materialize. As Margarita Balmaceda writes, control over key aspects of Ukraine's energy policy "was actually given to economic actors with a clear interest in the maintenance of Ukraine's energy dependency status quo."[5]

This chapter discusses four ways in which high levels of corruption in Ukraine's energy sector have ramifications upon transatlantic and European security. First, it leads to the prioritization of corruption and short term interests over reforms. Second, corruption is exported to western European NATO and EU members. Third, instability in energy relations between Russia and Ukraine could lead to a repeat of the 2006 and 2009 crisis crises. Fourth, prioritization of personal over national interests, as seen in the imprisonment of Julia Tymoshenko, derails Ukraine's European integration and leads to its possible geopolitical re-orientation towards Russia and the CIS.

---

[4] RosUkrEnergo was established by Presidents Vladimir Putin and Kuchma at a July 2004 Yalta meeting to replace Eural-Trans Gas, established in 2002.

[5] Margarita M.Balmaceda, *Energy Dependency, Politics and Corruption in the Former Soviet Union* (Abingdon and New York: Routledge, 2008), p. 76.

It's a Gas

Ukraine remains the seventh largest gas consumer in the world because gas consumption is highly inefficient and Ukraine's energy intensity is 2-2.5 times higher than EU levels. Around one-third is produced domestically with the remainder imported from Russia and Central Asia. Consumption of gas did not significantly decrease during the economic depression of the 1990s (when Ukraine suffered one of the highest declines in GDP in the former USSR) because a large proportion of the imported gas was re-exported. Re-export of gas brought huge profits for a large proportion of Ukraine's elites. Annual consumption of 75-78 bcm of gas remained consistent throughout the first two decades of Ukrainian independence with only the U.S. and Russia, both of whom have far larger populations and economies, consuming more gas. Following the 2004 Orange Revolution and Viktor Yushchenko's election as president of Ukraine, Russia massively increased the price of gas edging it towards "market levels" that Ukraine pays since 2010. The increase in imported gas prices has led to a halving of Ukraine's gas consumption to 35-40 bcm.

Ukrainian gas trader Ihor Bakay, in exile in Russia since the 2004 Orange Revolution, once famously said that most Ukrainian oligarchs made their initial capital on the re-export of Russian gas. Ukraine's private gas debts to Russia were passed to the Ukrainian state. Contracts "were most often with offshore companies of dubious provenance and involved the transfer of highly profitable areas of activity to them at the expense of state income and state decision-making power."[6] The Ukrainian state and taxpayers were lumbered with debts incurred from corrupt energy deals.

High gas dependency upon Russian energy supplies and a close corrupt relationship between Russian and Ukrainian elites in the energy sector has stymied Ukraine's Euro-Atlantic integration throughout the last two decades. From May 1997, when an agreement on the Black Sea Fleet was signed with Russia, until 2009 Ukraine was "paid" a paltry $100 million annual rent for the Sevastopol Black Sea Fleet naval base. Ukraine never received any of this rent because it was used to pay off gas debts accumulated by private traders and oligarchs

[6] Balmaceda, p. 73.

who had passed their debts to the Ukrainian state. The Black Sea Fleet has therefore been a hostage to Ukraine's energy dependency and corruption. The April 2010 "Kharkiv accords" extended the Sevastopol base until 2042-2047 in exchange for a mythical "thirty percent discount" on gas import prices. The Black Sea Fleet has de facto obtained a permanent naval base that, as in the case of Georgia during the 2008 war with Russia, could be used to promote separatism and block Ukraine's Euro-Atlantic integration.[7]

As Balmaceda points out, energy became both a "bottleneck in the country's economic development and Ukraine's Achilles Heel in its relations with Russia" that has led to multi-vector and muddled foreign policies that have balanced between the CIS and Europe. Dependency has led to "increased pressure for closer economic and political integration with Russia" and thereby increased Ukraine's vulnerability and reduced its ability to negotiate from a strong position.[8] Russian-Ukrainian energy corruption undermined rhetoric in favor of Euro-Atlantic integration by Presidents Kuchma and Yushchenko.

Russia has always controlled the narcotics fix that Ukraine, a "gas junky," has needed. Rising energy prices have forced some business owners to restructure Ukrainian industry, and Ukrainian companies have substituted the use of natural gas with more efficient technology, including the use of pulverised coal that requires no inputs of gas (Ukraine has the 9th largest coal deposits in the world). Nonetheless, Ukraine remains one of the top five world consumers of gas and Ukraine's big business continues to prefer Yanukovych's willingness to trade sovereignty for "cheaper" Russian gas (for example, through the 2010 extension of the Black Sea Fleet base). As with all Ukraine's gas contracts, the touted benefits of "discounted gas" were illusory except for a small group of corrupt elites and oligarchs.

As Roman Kupchinsky has pointed out, Ukraine is one of the most energy wasteful countries in the world. It consumes more natural gas—74 billion cubic meters in 2003—than Poland, Hungary, the Czech Republic, and Slovakia combined. Despite the huge amount of

---

[7] See T.Kuzio, *The Crimea: Europe's Next Flashpoint?* (Washington DC: The Jamestown Foundation, November 2010).

[8] Balmaceda, p. 77.

energy Ukraine consumes—1.5% of the world's total energy consumption according to the U.S. Energy Information Administration (EIA)—Ukraine's GDP of $300 billion in 2004 was far below Poland's figure of $463 billion.[9]

Households also benefitted, as Ukraine had one of the highest subsidized utility prices in the former USSR. There were no price increases of household utilities from 1998 until Russia dramatically increased gas prices following Yushchenko's election in December 2004. In 2006 Ukrainian households paid less for utilities than households in Belarus, or approximately twenty percent of the real cost of imported gas. The huge subsidy was passed to the state-owned Naftohaz Ukrainy gas company that made it perennially close to bankruptcy and added two percent to the state's budget deficit.

A condition of the 2008 and 2010 IMF agreements with the Tymoshenko and Nikolai Azarov governments respectively was the removal of subsidies to household utilities (and thereby reducing the budget deficit and transforming Naftohaz Ukrainy into a solvent entity). The Tymoshenko government ignored the IMF demand in the midst of the 2008-2010 global financial crisis and on the eve of January 2010 presidential elections. The Azarov government went half way and increased household utilities by fifty percent in August 2010 but then retreated and did not implement the second fifty percent increase, fearful of the consequences of public disapproval for the Party of Regions ahead of the October 2012 parliamentary elections.[10] The Azarov government has instead gone the traditional Ukrainian route of seeking cheaper below market Russian gas at the expense of national interests such as giving up control over its gas pipelines.

Energy Corruption

Every gas contract signed by Ukraine has been opaque and corrupt, benefitting Ukrainian elites rather than the state.[11] Balmaceda writes,

[9] Roman Kupchinsky, "Ukraine: An Unrepentant Gas Junkie," *RFERL Commentary*, 17 January 2006.

[10] See Roman Olearchyk, "Ukraine: time to call the IMF?" *Financial Times*, 3 October 2011.

[11] Balmaceda, p. 76.

"neither democrat nor oligarch, nationalist or friend of Russia seemed to be able to resist the temptation of its embrace or, perhaps, the fear of violent retribution reserved by the ultimate organizers of energy corruption for those who might seek to dismantle their profitable schemes."[12] Nationalists in the Yushchenko team were as willing to participate in energy corruption as pro-Russian politicians in the Kuchma or Yanukovych administrations. Retribution for not playing by the rules set out by elites has been tough against politicians such as Tymoshenko who has twice been imprisoned for seeking to dismantle schemes that provide energy rents.

Edward Chow,[13] Senior Fellow in the National Security Program at Washington's Center for Strategic and International Studies, believes Ukrainians successfully established one of the most efficient schemes in the world for extracting corrupt rents from the energy sector:

> *"If you set out to design a gas sector that is optimised for corruption, it might look very much like Ukraine's. You would have at its centre a wholly state-owned company that is not accountable to anyone other than the head of the country who appoints its management. It would operate non-transparently without discipline by either shareholders (who might demand legal rights as owners) or capital markets (since there is an implicit sovereign guarantee behind the company's borrowings). Domestic production would be priced artificially low, ostensibly for social reasons, leading to a large grey market on domestic gas supply that is allocated by privileged access rather than by price."*

Chow continues:

> *"As a result, consumers who have access to a cheap supply of gas use it in a wasteful manner and those who do not suffer shortages. Low gas prices suppress domestic gas production and energy efficiency, thereby necessitating the import of large volumes of gas which 'coincidently' is controlled by the same state monopoly. Even though the state company buys gas from another state monopoly*

[12]Balmaceda, p. 137.

[13]Interview with Edward Chow, Washington DC, June 2, 2011. E. Chow's biography can be found here: http://csis.org/expert/edward-c-chow.

from a neighbouring country, more frequently than not it utilises a middleman who has no discernible capabilities to import gas. The opaque middleman is paid handsomely in-kind, rather than with money, and in turn resells the gas to a higher-value market beyond the country's borders."

Finally, Chow believes:

"Foreign investment in the gas sector is discouraged since this would highlight the failings of the state company and force market-clearing domestic pricing, which would spoil the whole corruption optimisation scheme that you have set up. The game itself is ultimately unsustainable and highly costly to the country, but it is also high rewarding to insiders in the game."

Ukraine has utilized opaque gas intermediaries since the mid 1990s: Itera (1995-2001), Eural-Trans Gas (2002-2004) and RosUkrEnergo (2005-2009). President Putin, according to Balmaceda, proposed the replacement of Eural Trans Gas with RosUkrEnergo in July 2004 because he wanted to utilize it for geopolitical purposes to export greater volumes of gas to the EU. Dmitriy Medvedev, then Gazprom Chairman, became together with Putin major beneficiaries of the new gas intermediary.

On the Ukrainian side RosUkrEnergo was established by President Kuchma, Prime Minister Yanukovych and Naftohaz Ukrainy CEO Yuriy Boyko. Boyko is named on a document dated July 2004 as a member of RosUkrEnergo's key management committee.[14] In March 2010, Global Witness, a British NGO, raised concerns that Boyko had returned as Minister of Energy and Coal in the Nikolai Azarov government.[15] With the Kuchma era coming to a close and the likelihood of Yushchenko being elected Ukraine's elites needed to ensure their survival in the new era. Boyko established the Republican Party (RPU) that, after a brief existence, merged with the Party of Regions

[14]See "It's a Gas It's a Gas. Funny Business in the Turkmen-Ukraine Gas Trade" (London: Global Witness, July 25, 2006), available at: http://www.globalwitness.org/library/its-gas-funny-business-turkmen-ukraine-gas-trade.

[15]http://www.globalwitness.org/library/global-witness-concerned-choice-new-ukraine-energy-minister.

in 2006. Then Foreign Minister Konstantyn Gryshchenko joined the RPU and continued in government service under Yushchenko as deputy head of the National Security and Defence Council and Ambassador to Russia. The Azarov government returned him as Foreign Minister, a position which he has used to support criminal charges against Tymoshenko. Tymoshenko, he alleges, is supported by Russia and the EU because the 2009 gas crisis was beneficial to them—but not to Ukraine.[16]

The Ukrainian side was interested in extracting greater energy rents—not in geopolitics or the national interest. This was clearly seen in the decision by President Kuchma to not accept the Russian proposal that RosUkrEnergo be a joint venture of two state-owned gas companies: Gazprom and Naftohaz Ukrainy. The Ukrainian side instead opted to include two individuals—Dmitriy Firtash and Ivan Fursin—as owners of 45 and 5 percent respectively of RosUkrEnergo with Gazprom owning the remaining 50 percent. Ukrainian Prime Ministers Yuriy Yekhanurov (2005-2006) and Viktor Yanukovych (2006-2007) supported the structure of RosUkrEnergo and ruled out Naftohaz Ukrainy buying out Firtash and Fursin.[17] Billions of dollars that could have potentially been earned annually in gas transit to Ukraine and the EU would have ensured that Naftohaz Ukrainy would never be close to bankruptcy. These profits were instead diverted to oligarchs and senior Ukrainian officials.

Global Witness, a London-based NGO, commented:

> *"Naftohaz Ukrainy, as the state gas company of Ukraine, has a clear role to play in the Ukrainian gas market. Gazprom, as a supplier of gas, has a commercial interest in this market too. But what is the commercial rationale for involving RosUkrEnergo, a company with no track record in the gas industry? What service does RosUkrEnergo provide to the gas trade that cannot be provided by Naftohaz Ukrainy?"*

Global Witness continued:

[16]Interview with Konstantyn Gryshchenko in *Ukrayinska Pravda*, 28 September 2011.

[17]Balmaceda, p. 134. See also R. Kupchinsky, *Gazprom's European Web*, Jamestown Foundation special report, February 2009. www.jamestown.org.

> *"RosUkrEnergo made profits of over US$700 million in 2005. Meanwhile, Naftohaz Ukrainy has accrued debts of over US$500 million, mostly to RosUkrEnergo. Is it in Ukraine's benefit for Naftohaz to be so indebted to this private company, or for Ukraine to cede half of its domestic market to RosUkrEnergo via the new joint venture UkrGazEnergo?"*[18]

The pro-Western Yushchenko did not increase transparency in Ukraine's energy sector or ensure that energy policy began to serve the national interest. The 2006 gas contract, signed following the first of two major gas crises with Russia during his presidency, maintained the RosUkrEnergo intermediary in Ukraine's gas relationship with Russia. The 2006 contract, "hints of a personal interest in the agreements" as Yushchenko, "not only knew of the real owners behind RosUkrEnergo but may have received important benefits from them."[19] The Ukrainian delegation did not follow instructions from the Ministry Foreign Affairs or the government (Prime Minister Yekhanurov was excluded from the negotiation) but from Firtash and Naftohaz Ukrainy Deputy Chairman Oleh Voronin.

Ukrainian Centre for Economic and Political Studies expert Valeriy Chaly, Deputy Foreign Minister under Yushchenko, described the 2006 gas contract as Ukrainian diplomacy's "Pearl Harbor." The gas price remained unstable, transit fees were set for five years, a non-transparent pricing mechanism was put in place which did not foresee an increase to market prices, and the right to re-export gas was taken from Naftohaz Ukrainy and given to RosUkrEnergo. Re-export of gas is a lucrative business and RosUkrEnergo again diverted potential profits away from Ukraine's state gas company that was saddled with supplying the unprofitable housing market. The profitable distribution of gas to Ukrainian industry was given to a newly established company UkrHazEnergo, a joint venture between Gazprom and RosUkrEnergo.

---

[18] "New Ukrainian administration must answer key questions about Naftohaz Ukrainy and RosUkrEnergo," *Global Witness*, 25 July 2006. See http://www.globalwitness.org/library/new-ukrainian-administration-must-answer-key-questions-about-naftohaz-ukrainy-and

[19] Balmaceda, p. 128.

In winter 2008-2009 the second gas crisis lasted longer, seventeen days, during a bitterly cold winter. Although Europeans welcomed the gas contract concluded between Prime Ministers Tymoshenko and Putin, the negotiations were not transparent and it has therefore remained unclear why Tymoshenko agreed to some of its terms. On the positive side the contract removed RosUkrEnergo and thus for the first time Ukraine-Russia traded gas on a bilateral level without the use of opaque intermediaries. The contract also agreed to introduce the principle, for the first time, of gradually increasing gas prices to market levels—although unexplainably transit fees were not to grow to market levels.

On the negative side, "[t]he deal imposed a very high base price for gas of $450 per thousand cubic meters; a strict payment regime, which following any default could force Ukraine to pay for gas in advance; very high take-or-pay clauses that forced Ukraine to pay for gas it could not possibly use; and no ship-or-pay obligations on Gazprom (hence Gazprom, without penalty, could reduce shipments to Ukraine)."[20] These aspects of the 2009 gas crisis have led to criminal charges launched against Tymoshenko.

Why did Tymosenko agree to these terms? As gas negotiations between Ukraine-Russia have never been transparent we will never know the answer to this question. Tymoshenko was under intense pressure from Yushchenko (who lobbied for RosUkrEnergo to remain in place) and the EU (to end the 17 day gas cut off). The EU therefore would have welcomed any deal that was made. Speculation has suggested Tymoshenko agreed to the terms in exchange for Putin agreeing to remove RosUkrEnergo. Yushchenko and Yanukovych have alleged that Tymoshenko agreed to the terms in exchange for Putin's support in the 2010 elections and after agreeing to extend the agreement for the Sevastopol Black Sea Fleet base. Tymoshenko may have been seeking to score a double strike against Yanukovych ahead of the presidential elections by eliminating a major source of funding from the gas lobby and Putin's support.

---

[20]Alan Riley, "Corruption in Kiev and an E.U. Trade Pact," *The New York Times*, October 4, 2011.

The criminal charge against Tymoshenko for abuse of office under a 1962 Soviet criminal code is a charge that is tantamount to having taken the wrong political decision in the 2009 gas contract. Such a criminal article or charge would not take place in any democratic country with the rule of law (see below). There would be as much grounds for applying the same criminal charges against those who signed the 2006 gas contract ("Ukrainian diplomacy's Pearl Harbor") as that of the 2009 contract. The 2010 "Kharkiv Accords" and the decision made by Ukraine to establish a gas consortium over Ukraine's pipelines with Russia (60%), a German company with ties to Gazprom (20%) and Ukraine with the remaining 20%, that would bring Naftohaz Ukrainy under greater influence of Gazprom, will also be construed as a bad political decision by the opposition. To establish a consortium, parliament will have to change the February 2007 law that prevents the sale, lease, or rent of the pipelines. The law was lobbied by then opposition leader Tymoshenko and voted through by a massive majority of 430 deputies (out of 450)—including by the Party of Regions—although opposed by the Yanukovych government.[21]

Western Ukrainians and Energy

Firtash, although a Hungarian name, grew up in the Bukovina city of Chernivtsi and is Ukraine's only western Ukrainian oligarch. Firtash has never provided much information about his rise to the senior levels of Ukrainian business and politics. A U.S. Embassy cable from Kyiv following a December 2008 meeting between Firtash and U.S. Ambassador William Taylor revealed interesting details about his early life and how he entered business. In addition to Firtash, present at the meeting were consultant and AmCit Zev Furst, and Andras Knopp, the Hungarian-born number two at RosUkrEnergo.

Firtash, according to the U.S. diplomatic cable, "acknowledged ties to Russian organized crime figure Seymon Mogilevich, stating he needed Mogilevich's approval to get into business in the first place. He was adamant that he had not committed a single crime when building his business empire, and argued that outsiders still failed to under-

[21]Pavel Korduban, "Ukraine and Russia Prepare New Gas Agreement," *Eurasia Daily Monitor*, vol. 8 Issue: 178 (28 September 2011).

stand the period of lawlessness that reigned in Ukraine after the collapse of the Soviet Union."[22] Ties between organized crime and the energy sector in the former USSR had long been written about in Western specialist publications but these had always been the brunt of legal proceedings from oligarchs keen to whitewash their images.[23]

Western Ukrainians have played a prominent role in Ukraine's corrupt energy sector. These have included Congress of Ukrainian Nationalists (KUN) leader Oleksiy Ivchenko who was based in Turkmenistan in the 1990s and worked with Bakay in the Itera gas intermediary. Ivchenko headed Naftohaz Ukrainy in 2005-2006 and KUN was a member of Yushchenko's Our Ukraine bloc in the 2002 and 2006 elections. Ivchenko's gas business partner Bakai financed the 2000-2001 coalition of national democratic and centrist factions that backed the Yushchenko government.

Firtash had seen the writing on the wall after the second round of the 2004 elections on 21 November when the Orange Revolution derailed plans to install Yanukovych as president. Firtash became an important financial backer for Yushchenko's election campaign following a December 2004 meeting between them and Serhiy Levochkin, a senior adviser to President Kuchma. Yushchenko reached agreement with the energy lobby in his first year in office and supported RosUkrEnergo throughout his presidency. In summer 2005 the presidential secretariat ordered a halt to the pending arrest of former Naftohaz Ukrainy CEO Boyko by Security Service (SBU) Chairman Oleksandr Turchynov, Prime Minister Tymoshenko's right-hand man.[24] As Global Witness asked, "And what of the criminal investigation concerning RUE, launched by former Ukrainian State Security chief Aleksandr Turchynov in 2005? Why was this investigation never completed, and why does the office of the current SBU chief deny that an investigation ever took place?"[25]

---

[22]www.guardian.co.uk/world/us-embassy-cables-documents/182121

[23]Jane's Information Group and Radio Free Europe-Radio Liberty were forced to retract articles on Eural Trans Gas and remove them from their web sites.

[24]Balmaceda, pp. 123-124.

[25] "New Ukrainian administration must answer key questions about Naftohaz Ukrainy and RosUkrEnergo," *Global Witness*, 25 July 2006. See http://www.globalwitness.org/library/new-ukrainian-administration-must-answer-key-questions-about-naftohaz-ukrainy-and.

Firtash, like all Ukrainian oligarchs, is a pragmatist with no ideology and can therefore cooperate with all political groups if they do not intervene in his business affairs. In Ukraine all members of the establishment have been willing to cooperate with Firtash except Tymoshenko who has not been willing to play by their rules. Firtash and Yushchenko had close relations throughout his presidency and Fitash invested in many cultural and educational projects that the president supported. In February 2009 Firtash was awarded a state medal in gratitude for his financial donations.[26]

At the December 2008 meeting between Firtash and U.S. Ambassador Taylor:

> *"Firtash admitted that he has "loyally served" as an unofficial advisor to President Yushchenko during tense gas negotiations with Russia and political crises dating back to the Orange Revolution in 2004. He reported that he met with the Yushchenko at his dacha (cottage residence) three times in the last week at the President's request. He described himself as a close friend and confidante of the President—someone the President can trust totally."*[27]

While maintaining a cooperative relationship with President Yushchenko at the parliamentary level, Firtash supported Yanukovych and the Party of Regions, becoming an alternative source of funding to established Donetsk oligarchs such as Rinat Akhmetov. Firtash, Yushchenko and Yanukovych had a common interest in their contempt for Tymoshenko and worked together to prevent her election in 2010. Firtash thereby "strengthened the position of those wanting a closer relationship with Russia."[28] The gas lobby elbowed aside Donetsk oligarchs, such as Akhmetov, and established a commanding influence over the Party of Regions.

---

[26]Firtash invested in the re-building of Cossack encampments, such as Baturyn, and donated a large endowment to Lviv's Ukrainian Catholic University. Firtash provided $7 million for the launch of a Ukrainian cultural program at Britain's Cambridge University. On Firtash's funding see the *Kyiv Post*, 24 September 2010 and 19 May 2011.

[27]www.guardian.co.uk/world/us-embassy-cables-documents/182121.

[28]Balmaceda, p. 136.

This strategy enabled the gas lobby to come to power with Yanukovych in 2010 when its representatives received key appointments in the Yanukovych administration. Party of Regions deputy Nestor Shufrych confirmed that the RosUkrEnergo gas lobby, within which he includes Levochkin, controlled the Yanukovych administration.[29] Levochkin became head of the Presidential Administration, Valeriy Khoroshkovsky was promoted from First Deputy Chairman (a position he received from Yushchenko) to Chairman of the SBU, and Boyko was appointed Minister for Fuel and Coal. Khoroshkovsky ordered the SBU's Alpha anti-terrorist special forces to storm Naftohaz Ukrainy offices in retaliation for the elimination of RosUkrEnergo from the 2009 gas contract. Firtash revealed his contempt for Tymoshenko to U.S. Ambassador Taylor at the December 2008 meeting and, together with Khoroshkovsky, was able to take revenge against Tymoshenko after Yanukovych came to power—even at the cost of possibly undermining Ukraine's European integration.[30] Firtash and Khoroshkovsky are partners on Inter, Ukraine's most popular television channel, an important resource during elections as its main audience is in Russian-speaking eastern Ukraine.

Tymoshenko and Selective Use of Justice: The Gas Connection

Tymoshenko had participated in opaque gas schemes during the 1990s when she was CEO of United Energy Systems. But, after Tymoshenko was appointed Deputy Prime Minister in the Yushchenko government and twice Prime Minister during Yushchenko's presidency, she worked to close gas intermediaries. Among Ukraine's politicians only Tymoshenko has sought to remove gas intermediaries and battle corruption in the energy sector, and in retaliation the establishment has twice imprisoned her in 2001 and 2011.

Tymoshenko was imprisoned in January-February 2001 on charges related to her position as CEO of United Energy Systems in the 1990s. United Energy Systems operated with Itera and her political sponsor was Prime Minister Pavlo Lazaenko. In 1998 Tymoshenko

[29]Interview with Nestor Shufrych in *Segodnya*, July 13, 2011.

[30]www.guardian.co.uk/world/us-embassy-cables-documents/182121

entered parliament in Hromada (Community), a political party established by Lazarenko. A year later, Lazarenko's parliamentary immunity was removed and he fled abroad, eventually seeking diplomatic asylum in the U.S., which put him on trial for money laundering. In 2006 Lazarenko was sentenced to nine years in prison by a U.S. court for extortion, money-laundering through American banks and fraud. In 1999 Tymoshenko launched her own political party, Batkivshchina (Fatherland), becoming the main political party in the Tymoshenko bloc in the 2002, 2006 and 2007 elections. Batkivschina is an associate member of the center-right European People's Party which has been vocal in condemning in the European Parliament democratic regression in Ukraine under Yanukovych and selective use of justice against Tymoshenko.

The arrest of Tymoshenko on August 5, 2011 was tied to the authorities' fear that they were losing control of the trial in three ways. Firstly, they did not see the outcome they had planned for, namely, daily proceedings pointing to alleged guilt on Tymoshenko's part. Evidence provided by witnesses for the prosecution during the trial was neutral or even supportive of Tymoshenko's innocence, as in the case of testimony by Naftohaz Ukrainy CEO Oleh Dubyna. This is because, Oleksiy Krasnopyorov believes, "the instinct of self-preservation amongst the 'political class' has come to the fore."[31] Secondly, the trial proceedings had degenerated into what former Ambassador to Ukraine Steven Pifer described as a "farce."[32] Other Western figures pointed to how the authorities were desperate to find any charge that would "stick." Thirdly, and most importantly, live proceedings from the court room and intense questioning of witnesses by Tymoshenko began to publicly reveal too many details of the inner corrupt workings of the Yushchenko and Yanukovych administrations.

The first witness to be called who was interrogated about RosUkrEnergo was former Prime Minister and Our Ukraine leader Yekanurov who has always had an image of a center-right reformer loyal to Yushchenko. Yekhanurov's replies about RosUkrEnergo showed him to be disinterested in questions of corruption, market pricing or national security—three issues that the gas intermediary

[31] *Ukrayinska Pravda*, August 2, 2011.

[32] Steven Pifer, "Undemocratic values will isolate Ukraine," *Kyiv Post*, July 2, 2011.

had greatest influence over. In a lengthy interview Yekhanurov said the "Problem of RosUkrEnergo is thought up, it is the basis of business conflicts between two gas traders."[33] For Yekhanurov the only important factor in his decisions as prime minister was the lower price of gas offered by RosUkrEnergo. The Russian demand in 2006 for Ukraine to pay $230 per 1000 cu m. was "unacceptable" and the Yekhanurov government therefore agreed to RosUkrEnergo's proposal of $95. Yekhanurov said: "Before me was a price of $230 from Gazprom and $95 from RUE. Therefore I chose 95." After his court interrogation, Yekhanurov repeated, "What difference does it make who will be selling the gas? As long as this gas is delivered to Ukraine." "I believe that Ukraine would agree (to use RosUkrEnergo) today if the gas price was not as bad as the one that exists today."[34]

The choice was in fact a non-choice as Yekhanurov should have asked his advisers and SBU how it was possible for RosUkrEnergo to offer a price that was 2.5 times cheaper unless there the pricing formula was not transparent. Yekhanurov chose a non-market pricing formula that was opaque and never questioned how RosUkrEnergo—with a statutory fund of only $US20,000—could be involved in gas trading in a business that had an annual turnover of billions of dollars.

One of the most damning witnesses in the Tymoshenko trial proved to be Dubyna, CEO of Naftohaz Ukrainy during the 2007-2010 Tymoshenko government. His testimony was not always favorable to Tymoshenko but was startling for his revelations about RosUkrEnergo's links to President Yushchenko whom, he argued, lobbied its interests during the 2008-2009 gas negotiations with Russia. As part of this lobbying effort, President Yushchenko and RosUkrEnergo sabotaged the negotiations Tymoshenko was conducting towards a lower gas price. Dubyna testified that he was called back to Kyiv on December 31, 2008 by President Yushchenko to prevent the signing of a cheaper price gas deal. Asked during his testimony if he had contact with RosUkrEnergo's founders and managers, Dubyna replied: "Yes. I met with them only in the presidential administration. Three times in Viktor [Yushchenko]'s office, twice in the office of [Viktor] Baloga," who was then chief of staff. Baloga resigned from the position of head

[33]Interview with Yuriy Yekhanurov in *Ukrayinska Pravda*, July 14, 2011.

[34]Ibid.

of the presidential secretariat in May 2009 and worked for Yanukovych's election campaign. Trans-Carpathia, Baloga's home region, was the only western Ukrainian region which voted for Yanukovych in the 2010 elections. Baloga was rewarded with the position of Minister of Emergency Situations in the Azarov government.

Political analyst Viktor Nebozhenko pointed out that Tymoshenko's arrest on August 5, 2011, which damaged negotiations towards, and future ratification of, a Deep Comprehensive Free Trade Agreement between Ukraine and the EU, coincided with the approaching interrogation of former President Yushchenko. "I believe these two events are inter-dependent," he said.[35] Mustafa Nayem in a lengthy analysis of Yushchenko's place in the trial wrote how testimony given at the trial had revealed concrete facts, "that confirm ties between the ex-president with the gas intermediary and the role these ties played in the 2009 gas crisis."[36] Yushchenko's testimony to the Tymoshenko trial supported the charges laid against her.

Dubyn revealed that up to December 31, 2008 Ukraine could have signed a contract with a price of $235 per 1000 cu.m. and gas transit prices of $1.8 which he believed were a good deal for Ukraine. But President Yushchenko forbade him from signing this favorably priced contract. Gazprom head Alexey Miller told his Ukrainian counterpart, Naftohaz Ukrainy head Dubyn: "I have a letter from RosUkrEnergo that they are ready to pay 295 for all the gas. This is backed by Yushchenko's word."[37] Investigative journalist Nayem concluded that this testimony will enter "the handbooks on questions of corruption in the highest levels of state officials."

Tymoshenko was first arrested on August 5, 2001 for "contempt of court" and sentenced on October 11, 2011 to seven years imprisonment, a three year ban from political life and fine of 1.5 billion *hryvnia* (approximately $200 million) The "7+3" charges ban Tymoshenko from the next two presidential and three parliamentary elections, conveniently removing her from politics during two potential presidential

[35]Viktor Nebozhenko cited on www.politdumka.kiev.ua, August 5, 2011.

[36]Mustafa Nayem, "Sud nad Yuliyu Tymoshenko: na arenu vykhiodyt Viktor Yushchenko," *Ukrayinska Pravda*, August 1, 2011.

[37]*Ukrayinska Pravda*, August 1, 2011.

terms for Yanukovych up to 2020. The sentence, refusal to heed the flood of Western criticism and new charges against Tymoshenko led the EU to cancel the October 20, 2011 Brussels visit by Yanukovych, where he was to finalize negotiations on the Association Agreement ahead if its signing in December 2011 at the EU-Ukraine summit in Kyiv.

To add oil to the fire, the Security Service (SBU) launched two additional criminal charges against Tymoshenko related to when she was CEO of United Energy Systems in 1995-1997 and the assassination of Viktor Yushchenko's protégé, banker Vadym Hetman in 1998. Both charges are linked to former Prime Minister Pavlo Lazarenko, who was sentenced in August 2006 by the U.S. to nine years imprisonment on money laundering charges.

SBU Chairman Valeriy Khoroshkovsky's intervention was the latest in a number of similar steps that all had the objective of damaging Ukraine's European integration. Khoroshkovsky was promoted to the position of First Deputy Chairman of the SBU by President Yushchenko in January 2009 and became its Chairman in March 2010. His interventions have led Ukrainian experts to conclude, as one of four members of the pro-Russian gas lobby, that he heads a pro-Russian group within the Yanukovych administration with the strategic objective of integrating Ukraine into the CIS Customs Union. Khoroshkovsky and Firtash jointly own Inter, Ukraine's most popular television channel, that is mainly watched in eastern-southern Ukraine, the voter base for the Party of Regions.

Tymoshenko's trial descended into what former U.S. Ambassadors to Ukraine Steven Pifer and William Taylor described as a "farce." Taylor described criminal prosecution of the opposition as a "bad precedent" and "madness", telling Radio Svoboda, Radio Liberty's Ukrainian language service, "It's hard to understand why criminal charges are brought against the political decisions of the previous administration" because it "is a very bad precedent for future governments." This is because "The next government may start judging the current one."[38]

[38]"Ex-US ambassador calls Yanukovych's policy 'madness'," *Interfax-Ukraine*, August 29, 2011, Steven Pifer, "Undemocratic Values will isolate Ukraine," and "Does Yanukovych get it?" *Kyiv Post*, July 28 and September 20, 2011.

The trial was viewed as a "farce" because Tymoshenko was sentenced under article 365 for abuse of office" which first was introduced in the 1962 criminal code when Nikita Khrushchev was Soviet leader. The article remained in Ukraine's criminal code that was adopted in 2001. No similar article exists in any rule of law-based European or U.S. democracy, as it punishes politicians for undertaking the "wrong" decision, a very subjective charge that is open to high degrees of political manipulation by those who are in power.

The criminal charge opens up a Pandora's box of counter-charges if the Yanukovych team is voted out of office, a threat that could make Yanukovych seek to hold on to power indefinitely. This, in turn, makes the likelihood of free elections in Ukraine in the near future a remote possibility.[39] The OSCE cannot recognize elections as having been held in "accordance with democratic values" if opposition leaders sit in jail. The sentencing of Tymoshenko and unwillingness to heed Western demands has placed the Association Agreement in jeopardy; even if it were to be signed there is no likelihood of the Agreement being ratified by the European Parliament and the parliaments of 27 EU members.

Ramifications for Transatlantic and European Security

High levels of corruption in Ukraine's energy sector have ramifications upon transatlantic and European security in four ways. First, the temptation of drawing high levels of rents from the energy sector has made Ukrainian elites prioritize short-term gain over the medium-long term benefits of reforms. Ukrainian elites have paid lip service to support for reforms throughout the last two decades and have failed to fulfil the demands placed upon them by the IMF and EU. Second, the inefficiency of Ukraine's energy sector, lack of modernization and corruption has external ramifications. Corruption and a poor business environment have made foreign investors wary of investing in Ukraine's energy sector and pipelines. Much of the corrupt proceeds from Ukraine's energy sector are exported to western European EU and NATO members. Offshore zones Cyprus and the British Virgin

[39]See T.Kuzio, "Can Ukraine Hold Free Elections Next Year?" *Eurasia Daily Monitor*, Volume: 8, Issue: 191 (October 18, 2011).

Islands are the first and fifth largest foreign investors in Ukraine. London and Britain have received a huge injection of finances from Ukraine, Russia and the CIS. Ukrainian oligarchs Renat Akhmetov and Viktor Pinchuk bought two of the most expensive properties in British history in 2009-2011 for a combined total of US$400mn.

Third, unstable energy relations between Ukraine and Russia could lead to further gas crises as took place in 2006 and 2009. As Ukraine is a major transit route for Russian and Central Asian gas to NATO and EU members, gas crises can lead to major security threats, as seen in the seventeen-day cut off in January 2009. Fourth, personal and business interests are of greater importance than the national interests of Ukraine which can derail Ukraine's integration into Europe and change the balance of power by leading to Kyiv's reorientation towards Russia and the CIS Customs Union. Revenge against Tymoshenko for removing the opaque gas intermediary RosUkrEnergo from the 2009 gas contract with Russia is more important in the eyes of Ukraine's elites than European integration. The signing of an Association Agreement (which includes a Visa Free Regime and a Deep and Comprehensive Free Trade Agreement) with the EU is far less important than personal revenge or enrichment—even though it would benefit the country and its citizens.

Conclusions

After two decades of Ukraine as an independent state, corruption has become an integral part of Ukrainian society, business and politics. Corruption emerged in the 1990s transition to a market economy and has never been tackled by any Ukrainian president. The energy sector has been the most tempting for corrupt elites from all political groups and regions as the greatest rents can be extracted from it that give incumbents greater advantage in politics. Western Ukrainians have played a prominent role in gas corruption and Ukrainian nationalist party leaders have dropped their anti-Russian rhetoric when travelling to Moscow to negotiate corrupt energy deals.

What continues to remain unclear is why Prime Minister Yushchenko supported Tymoshenko's policies to reduce energy corruption while President Yushchenko blocked both Tymoshenko governments' attempts to undertake similar policies. President Yushchenko continued the state and executive's involvement in corruption that he had opposed as prime minister. Energy corruption was a major factor in bringing about the failure of Yushchenko's presidency and the Orange Revolution and thereby facilitated the coming to power of Yanukovych and the gas lobby.

Ukraine's mismanaged energy policy and high level corruption in the energy sector has four ramifications for transatlantic and European security. First, it leads to vacuous rhetoric about reforms and a failure to fulfil demands to reform sectors of the economy, finances, rule of law and corruption that are placed upon Ukraine by the IMF and EU. Second, it leads to the export of corruption to western European members of NATO and the EU. Third, unstable relations with Russia could lead to further gas crises as shook NATO and EU members in 2006 and 2009. Finally, the domination of personal over national interests, as seen in the revenge undertaken by the Yanukovych administration against Tymoshenko for successfully removing RosUkrEnergo, can change the balance of power in Europe. The imprisonment of Tymoshenko was strongly condemned by the EU, US and Canada and will lead to the European Parliament and 27 EU members failing to ratify the Association Agreement. If European integration is blocked, Kyiv has threatened to join the CIS Customs Union, which would establish a second 'Belarus' on the eastern borders of NATO and the EU.[40]

[40]See T.Kuzio, "Surreal Eastern Partnership Summit: EU Gives Ukraine Last Red Card," *Eurasia Daily Monitor* Volume: 8 Issue: 185 (October 7, 2011).

About the Authors

David Koranyi is a Non-Resident Fellow at the Johns Hopkins SAIS Center for Transatlantic Relations. His research is on central European and transatlantic energy security issues. He served as the chief foreign policy and national security advisor to the Prime Minister of the Republic of Hungary, Gordon Bajnai, in 2009-2010. He worked in the European Parliament as chief foreign policy advisor and head of cabinet of Hungarian MEP Csaba Tabajdi between 2004-2009. Previously he was a political advisor at the Hungarian National Assembly and a junior researcher at GKI Economic Research Institute, a Hungarian research firm. He pursued undergraduate studies in political economy and business administration then obtained a Master's degree in International Relations and Economics, with a major in Foreign Affairs. He also studied law for 8 semesters. He has published articles and essays on energy security, Hungarian and U.S. foreign policy, European integration, the western Balkans and the United Nations. He is a regular speaker at international conferences on foreign and security policy and energy security. He was a member of the Hungarian NATO Strategic Concept Special Advisory Group and a recipient of the German Marshall Fund's 2010 Marshall Memorial Fellowship.

Reinis Aboltins is a European policy researcher specializing in energy policy at the Centre for Public Policy PROVIDUS, which he joined at the end of 2009. He has written on national, regional and EU energy policy, including, but not limited to energy policy planning, outstanding energy policy issues, European energy legislation, energy islands in the EU as well as climate issues. Prior to his current position he spent four years as the public affairs director of Baltic Communication Partners, a private public relations and public affairs company. During this period he developed his specialization in energy policy, energy security and EU-Russia relations. He initiated and produced a first-of-its-kind bi-monthly electronic journal on energy issues, supported by Latvia's national energy incumbent *Latvenergo*. Over the course of four years he wrote more than 80 articles and reviews on energy issues in Latvia, the Baltic states, the EU and other regions of the world. Previously he worked for the UNDP Latvia; and served as adviser to the State President as well as to the Head of the

Naturalization Board, Director of Social Integration Department at the Latvian Ministry of Justice. He is co-author of the energy chapter of the Latvian national long-term sustainable development strategy *Latvia 2030* and also co-author of civil society chapter of the framework document for the social integration–national program *Integration of Society in Latvia*. He is a frequent speaker and moderator at local and international events on energy, climate, sustainable development and human rights. He holds a European Masters degree in Human Rights and Democratization from the University of Padova in Italy and the Raoul Wallenberg Institute of Human Rights at the Lund University in Sweden, as well as a BA in political science from the University of Latvia.

Kornel Andzsans-Balogh is an energy market researcher who works as Research Associate at the Regional Center for Energy Policy Research (REKK) at the Corvinus University of Budapest. He conducted market analyses of the central and southeastern European gas markets and participated in the REKK's central and southeastern European security of supply project, where his research focus was mid- and long-run security of supply assessment and developing regulatory risk measurement tools. His interests are eastern European natural gas markets and CIS producers. Currently he is conducting research on economic incentives of biomethane injection into natural gas grids. The research grant is provided by the German Federal Environmental Foundation (Deutsche Bundesstiftung Umwelt—DBU). He holds an MA in Economics from the Corvinus University of Budapest.

Pál Belényesi is a competition lawyer with a history of working for the European Commission and national competition authorities. He currently leads a strategic consulting firm in Hungary, while teaching part time at the Florence School of Regulation at the European University Institute. In the past he was a visiting lecturer at the University of Verona and the University of Debrecen. He received his law degree from the University of Debrecen, specialized at the University of Antwerp, and completed his studies with an LL.M. at the European University Institute in Florence. He is the author of several academic articles, monographs, and a co-author of a book. He will complete his PhD in 2012 on water pricing and environmental regulation.

Mihaela Carstei is the Associate Director of the Energy and Environment Program at the Atlantic Council, specializing in infrastructure development, bilateral energy relations, emerging energy technology, and policy related to energy and the environment. Prior to joining the Atlantic Council, she conducted research on peace education as a graduate assistant to Syracuse University Professor Peter Castro; and on U.S. economic, foreign, and energy policies for Syracuse University Professor Stuart Brown. Previously, she developed research reports for the Montana World Trade Center at the University of Montana. She is a PhD. candidate in International Economics at the Academy of Economic Studies in Bucharest, Romania. She completed her M.A. at Syracuse University's Maxwell School of Citizenship and Public Affairs with a concentration in International Environmental Policy and Global Markets. She holds a B.S. in International Business and Management from the University of Montana, and a B.A. in Cybernetics and Economic Forecast from the Academy of Economic Studies. She was a fellow of the Soros Foundation/Open Society Institute Program in 2002-2003.

András Deák is a Research Director at the Center for EU Enlargement Studies, Central European University. Between 2007 and 2009 he served as Executive Director of the Hungarian Institute of International Affairs. From 1998 he was a researcher in charge of post-Soviet affairs at the Teleki László Institiute, Budapest. He has held several seminars on Contemprorary Russian Politics and International Energy Policy at the Budapest Corvinus University. He received his MA and Ph.D. in International Relations from the University of Economic Sciences in Hungary.

Nikolas Foster is a Communication Specialist at the Pacific Northwest National Laboratory (PNNL) in Richland, Washington, one of ten U.S. Department of Energy (DOE) national laboratories managed by DOE's Office of Science. At PNNL, he supports the Electricity Infrastructure Division within the Energy and Environment Directorate. His work focuses on promoting PNNL's grid-related research. He is the communications lead for the Pacific Northwest Smart Grid Demonstration Project, the largest regional smart grid demonstration in the United States. Previously, he worked as a consultant for the World Bank's Energy Sector Management Assistant Program (ESMAP). At ESMAP, he worked on researching best practices in

Low-energy Building Standards for the Energy Efficient Cities Initiative Program (EECI). He has an M.A. in Energy and Environmental Policy and International Economics from the Paul H. Nitze School of Advanced International Studies (SAIS) in Washington, DC. While at SAIS, Nikolas worked at the Center for Transatlantic Relations as a research analyst supporting Director Dan Hamilton on the *Transatlantic Economy* series, as well as *Europe 2020: Competitive Or Complacent?* Together with Hamilton he authored the chapter "The Obama Administration and Europe," in the EUISS publication *The Obama Moment—European and American perspectives* (2009).

Tamás Kenessey is a lawyer working as the head of cabinet of a Hungarian Member of the European Parliament. He is responsible for a wide variety of subjects including energy, environmental policy, climate change, management of natural resources and external energy policy in EU-China and EU-Russia relations. Previously he worked as the secretary general of a nine-strong political group in the European Parliament, responsible for horizontal coordination of legislative activity. He was involved in the political and legislative process leading to the adoption of the EU's energy and climate package. He is currently responsible for the political and administrative management of two research projects mandated by the European Parliament, one on the methanol economy and the other on the sustainable use of natural resources. He obtained a Masters Degree in Law at the Eötvös Lóránd University, Faculty of Law, Budapest, with a special focus on international public and civil law, European community law and public administration.

Taras Kuzio is a Non-Resident Fellow at the Center for Transatlantic Relations of the Paul H. Nitze School of Advanced International Studies, John Hopkins University. Previously, he was an inaugural Austrian Marshall Plan Foundation Fellow at CTR. He has been a Visiting Professor at the Institute for European, Russian and Eurasian Studies, Elliott School of International Affairs, George Washington University and Senior Research Fellow in the Centre for Russian and East European Studies at the University of Birmingham. His previous positions have included a Resident Fellow at the Centre for Russian and East European Studies, University of Toronto, and Adjunct Research Professor, Institute for European and Russian Studies, Carleton University, Ottawa. He also served as Head of Mission of the NATO Information and Documentation Centre in Kyiv, Ukraine. He is the author and editor of 14 books, including *Theoretical and Comparative Perspectives on Nationalism* and *Ukraine-Crimea-Russia: Triangle of Conflict*, both published in the *Soviet and Post-Soviet Politics and Society series* by Ibidem-Verlag, 2007. He is the author of 5 think tank monographs and 25 book chapters, the most recent being 'Civic Nationalism and the Nation-State: Towards a Dynamic Model of Convergence' in Ireneusz P. Karolewski and Andrzej Marcin Suszycki eds., *Nationalism in Contemporary Europe* (Banham, MD: Roman and Littlefield, Lexington Books, 2009) and 'Ukraine. Muddling Along' in Sharon L. Wolchik and Jane L. Curry eds., *Central and East European Politics: From Communism to Democracy, Second Edition* (Banham, MD: Roman and Littlefield, 2011). He has authored over 60 scholarly articles on post-communist politics, Ukraine, and nationalism and has guest edited 6 special issues of academic journals, including a recent issue of *Communist and Post-Communist Studies* on 'Communist Successor Parties in Central-Eastern Europe and Eurasia'. He is the editor of the bi-monthly *Ukraine Analyst*. He received a BA in Economics from the University of Sussex, an MA in Soviet Studies from the University of London and a PhD in Political Science from the University of Birmingham, England. He was a Post-Doctoral Fellow at Yale University.

Maximilian Kuhn is a research associate at the Center for Energy, Marine Transportation and Public Policy (CEMTPP) at the School of International and Public Affairs (SIPA), Columbia University, where he coordinates the EURASIA GAS 2030 Scenario Project. He also holds a visiting research associate position and is the chief editor of the EUCERS Strategy Papers at the European Centre for Energy and Resource Security (EUCERS), at the Department of War Studies, King's College London, as well as doctoral student at the Freie Universität Berlin, Germany, in the PhD program for *Caspian Region Environmental and Energy Studies (CREES)*. Prior to commencing his PhD, he worked in public affairs and institutional public relations, corporate PR and agenda setting. His current research focuses on energy and energy policy decision-making and future studies, such as scenarios for the Eurasian gas market for 2030 and the future of the Iranian natural gas sector. His academic work on the Eurasian gas market, the liberalization of the Turkish energy market and on unconventional gas has been published in specialized journals, and he has contributed to and has been quoted in international media such as BBC World Service, *Der Spiegel*, European Energy Review and other media outlets specializing on energy.

Mark Olsthoorn is a visiting scholar from the Netherlands at the University of Maryland Center for Integrative Environmental Research. His research addresses cross-cutting public policy problems at the regional to international level related to energy and climate change impacts and mitigation. He holds a Masters degree in Aerospace Engineering from Delft University of Technology. Before coming to Maryland, Mr. Olsthoorn has worked at IMSA Amsterdam, an entrepreneurial consultancy for sustainability and innovation in Amsterdam, the Netherlands.

Michael Stanton-Geddes works in the energy and resources sector, providing economic and policy consultant services to institutions including the World Bank and multinational corporations. He worked in Washington, DC and Brussels, Belgium for Fleishman-Hillard, and conducted research in Morocco and the United States at the Center for Transatlantic Relations. He is a graduate of Tufts University and the Johns Hopkins University School of Advanced International Studies (SAIS).

Frank Umbach is Senior Associate and Head of the International Energy Security Program at the Centre for European Security Strategies (CESS) GmbH, Munich-Berlin & Associate Director of the European Centre for Energy and Resource Security *(EUCERS)*, Department of War Studies, King's College London. He is an expert for energy, foreign and security policies with a regional expertise on Russia, the Caspian Region, Asia-Pacific (i.e. China and Japan) and Europe. Previously, he was head of the International Energy Security and Security Policies in Asia-Pacific Programs at the German Council on Foreign Relations (DGAP) in Bonn and Berlin; and worked before at the Federal Institute for East European and International Studies (BIOst) in Cologne; in the office of the NATO General Secretary in Brussels; and the Japan Institute for International Affairs, the institute of the Japanese Foreign Ministry. He has conducted consultancy work for the German Ministries of Foreign Affairs and Defense; European Commission and European Parliament, U.S. Departments of State and Energy, U.S.-China Economic and Security Review Commission of the U.S. Congress, the Lithuanian Government, the Polish Foreign Ministry, the British House of Lords, NATO and the OSCE. He studied Political Science, East European History and European/International Law at the Universities of Marburg and Bonn. He is the author of more than 300 publications in 25 countries, he also publishes regularly for the "Geopolitical Information Service" (www.geopolitical-info.com). He co-authored the study "Strategic Perspectives of Unconventional Gas in Europe: A Game Changer with Implications for the EU's Common Energy Security Policies", EUCERS-Strategy Papers, King's College, Vol. 1, No. 1, May 2011.

Kirsten Westphal is based at the Stiftung Wissenschaft und Politik (SWP), the German Institute for International and Security Affairs in Berlin, Germany, where she works in the section on International Energy Relations and Global Energy Security. Before, she was Assistant Professor for International Relations at the Department for Political Science at the Justus Liebig University in Giessen, Germany. She has been a consultant and scholarship holder in the energy industry and has worked in several EU and OSCE missions in Latin America, Eastern Europe, the CIS and Asia. She has published widely on international energy relations and EU external energy relations. Recent publications include *Global Energy Governance in a Multipolar World* (with Dries Lesage and Thijs van de Graaf); *The Energy Charter Treaty Revisited* SWP Comments 2011; *Internationale Energiebeziehungen (International Energy Relations)*, in: Wichard Woyke (ed.), *Handwörterbuch Internationale Politik*; *Solar Power from North Africa* (with Isabelle Werenfels) SWP Research Paper 2010; *Energy Policy in the USA* (with Stormy-Annika Mildner) in: World Energy Council, Energy for Germany 2010, p.16-44; and *Russian Gas, Ukrainian Pipelines and European Supply Security*, SWP Research Paper 2009.